CEOE
Field 14

OSAT
Physics
Teacher Certification Exam

By: Sharon Wynne, M.S
Southern Connecticut State University

"And, while there's no reason yet to panic, I think it's only prudent that we make preparations to panic."

XAMonline, INC.
Boston

Library of Congress Cataloging-in-Publication Data

Wynne, Sharon A.
    OSAT Physics Field 14: Teacher Certification / Sharon A. Wynne. -2nd ed.
    ISBN 978-1-58197-663-2
    1. OSAT Physics Field 14.   2. Study Guides.   3. CEOE
    4. Teachers' Certification & Licensure.   5. Careers

Disclaimer:

The opinions expressed in this publication are the sole works of XAMonline and were created independently from the National Education Association, Educational Testing Service, or any State Department of Education, National Evaluation Systems or other testing affiliates.

Between the time of publication and printing, state specific standards as well as testing formats and website information may change that is not included in part or in whole within this product. Sample test questions are developed by XAMonline and reflect similar content as on real tests; however, they are not former tests. XAMonline assembles content that aligns with state standards but makes no claims nor guarantees teacher candidates a passing score. Numerical scores are determined by testing companies such as NES or ETS and then are compared with individual state standards. A passing score varies from state to state.

Printed in the United States of America   œ-1

CEOE: OSAT Physics Field 14
ISBN: 978-1-58197-663-2

Table of Contents

SUBAREA II.   **MECHANICS AND THERMODYNAMICS**

Great Study and Testing Tips!

What to study in order to prepare for the subject assessments is the focus of this study guide but equally important is *how* you study.

You can increase your chances of truly mastering the information by taking some simple, but effective steps.
Study Tips:

1. <u>Some foods aid the learning process.</u> Foods such as milk, nuts, seeds, rice, and oats help your study efforts by releasing natural memory enhancers called CCKs (*cholecystokinin*) composed of *tryptophan*, *choline*, and *phenylalanine*. All of these chemicals enhance the neurotransmitters associated with memory. Before studying, try a light, protein-rich meal of eggs, turkey, and fish. All of these foods release the memory enhancing chemicals. The better the connections, the more you comprehend.

Likewise, before you take a test, stick to a light snack of energy boosting and relaxing foods. A glass of milk, a piece of fruit, or some peanuts all release various memory-boosting chemicals and help you to relax and focus on the subject at hand.

2. <u>Learn to take great notes.</u> A by-product of our modern culture is that we have grown accustomed to getting our information in short doses (i.e. TV news sound bites or USA Today style newspaper articles.)

Consequently, we've subconsciously trained ourselves to assimilate information better in <u>neat little packages</u>. If your notes are scrawled all over the paper, it fragments the flow of the information. Strive for clarity. Newspapers use a standard format to achieve clarity. Your notes can be much clearer through use of proper formatting. A very effective format is called the *<u>"Cornell Method."</u>*

Take a sheet of loose-leaf lined notebook paper and draw a line all the way down the paper about 1-2" from the left-hand edge.

Draw another line across the width of the paper about 1-2" up from the bottom. Repeat this process on the reverse side of the page.

Look at the highly effective result. You have ample room for notes, a left hand margin for special emphasis items or inserting supplementary data from the textbook, a large area at the bottom for a brief summary, and a little rectangular space for just about anything you want.

3. **Get the concept then the details.** Too often we focus on the details and don't gather an understanding of the concept. However, if you simply memorize only dates, places, or names, you may well miss the whole point of the subject. A key way to understand things is to put them in your own words. If you are working from a textbook, automatically summarize each paragraph in your mind. If you are outlining text, don't simply copy the author's words.

Rephrase them in your own words. You remember your own thoughts and words much better than someone else's, and subconsciously tend to associate the important details to the core concepts.

4. **Ask Why?** Pull apart written material paragraph by paragraph and don't forget the captions under the illustrations.

Example: If the heading is "Stream Erosion", flip it around to read "Why do streams erode?" Then answer the questions.

If you train your mind to think in a series of questions and answers, not only will you learn more, but it also helps to lessen the test anxiety because you are used to answering questions.

5. **Read for reinforcement and future needs.** Even if you only have 10 minutes, put your notes or a book in your hand. Your mind is similar to a computer; you have to input data in order to have it processed. *By reading, you are creating the neural connections for future retrieval.* The more times you read something, the more you reinforce the learning of ideas.

Even if you don't fully understand something on the first pass, *your mind stores much of the material for later recall.*

6. **Relax to learn so go into exile.** Our bodies respond to an inner clock called biorhythms. Burning the midnight oil works well for some people, but not everyone.

If possible, set aside a particular place to study that is free of distractions. Shut off the television, cell phone, pager and exile your friends and family during your study period.

If you really are bothered by silence, try background music. Light classical music at a low volume has been shown to aid in concentration over other types.

Music that evokes pleasant emotions without lyrics are highly suggested. Try just about anything by Mozart. It relaxes you.

7. <u>Use arrows not highlighters.</u> At best, it's difficult to read a page full of yellow, pink, blue, and green streaks.

Try staring at a neon sign for a while and you'll soon see my point, the horde of colors obscure the message.

8. <u>Budget your study time</u>. Although you shouldn't ignore any of the material, *allocate your available study time in the same ratio that topics may appear on the test.*

COMPETENCY 1.0   Understand the relationships and common themes that connect mathematics, science, and technology

Skill 1.1   Apply the laws of physics to various sciences and other disciplines outside of physics

The laws of physics are those that govern all parts of the natural world. The laws of physics reveal the root explanation for every geological, chemical, biological, and astronomical phenomenon. Let's examine how knowledge of the different aspects of physics is useful in understanding phenomenon from each field.

Geology: The study of the physical laws that affect the earth and its processes is known as geophysics. Thermodynamics is needed to understand how heat moves between the core, mantle, and crust of the earth. Additionally, thermodynamics, along with the study of electromagnetism, is needed to comprehend and predict meteorological events. Mechanics is required to analyze the movement of the tectonic plates, earthquakes, and other seismic activity. Mechanics, particularly those of waves, is also key in the study of hydrology.

Chemistry: Physical chemistry is the branch of chemistry concerned with physical behavior of molecules and the physical aspects of chemical reactions. Concepts from modern physics including nuclear physics and quantum and statistical mechanics are of key importance in this field. Thermodynamics and basic mechanics are also useful in understanding the progress of chemical reactions. Chemists have also made extensive use of the laws of electromagnetism in designing analysis tools such as NMR spectrometers.

Biology: The application of physical theories and methods in biological systems is known as biophysics. This field is largely devoted to the micromechanics of the cell, particularly the cytoskeleton and the physical structure of proteins. However, understanding larger biological systems also requires an understanding of physics. For instance, thermodynamics is needed to analyze heat transfer in warm and cold blooded animals. Mechanics is necessary to understand and develop therapies for the musculoskeletal system.

Astronomy: Astrophysics is the study of the physics of the universe. Mechanics is important in predicting the motion of planets, comets, and other celestial bodies. Thermodynamics is used in studying the transfer of energy and the heating and cooling of planets and stars. Optics and an understanding of sound and light are also necessary, particularly for the proper use of telescopes and other instruments used for studying the universe.

Skill 1.2   Analyze the use of physics, mathematics, and other sciences in the design of a solution to a given scientific or technological problem

Technology includes all types of devices and methods created by humans to achieve a given objective. Simple tools, complex machines, manufacturing processes, and software are all types of technology. It is true that scientific reasoning and a grasp of scientific laws are not essential in the development of new technology. Indeed, our early ancestors devised many primitive technologies without the aid of science as we understand it. Various more modern inventions have also been created with little or no scientific assistance. However, as inventions become more complex, a firm grasp of scientific principles is needed to develop effective solutions for problems. Additionally, the development of new technology can be rapidly accelerated by taking advantages of scientific concepts.

Firstly, science can help us in identifying problems and suggesting solutions. For example, molecular biology and biochemistry are useful in helping us identify underlying problems in clinical conditions and discovering potential targets for pharmaceuticals. Similarly, various atmospheric and chemical measurements have been used to quantify climate change and suggest mechanisms. Additionally, a study of organisms can show us how evolution has solved problems. One classic example of this is the invention of Velcro which mimics the action of burrs on animal fur.

Second, an understanding of a broad range of sciences may be helpful in predicting the suitability of various options and in the creation of new ones. Consider, for instance, an implantable orthopedic device such as a hip replacement that must both support weight and be non-toxic. We can use mechanics to predict the forces that the device must withstand. Similarly, principles of material science help us determine what type of substance would have appropriate mechanical and chemical properties. If current substances do not meet the requirements, organic chemistry may allow the creation of new ones.

Finally, scientific reasoning can assist in the process of testing and refining a new technical design. The basics of the scientific method allow us to determine what tests should be done and what we can learn from each. Mathematics allows us to model the behavior of the technology. Statistics is also particularly useful in helping us determine how confident we can be in the results of the testing and whether additional experimentation is necessary. It may also help us quickly hone in on optimal conditions, for instance, in a new manufacturing process with multiple variables.

Skill 1.3   Analyze the role of technology in the advancement of scientific knowledge

Scientific discoveries are important in the development of new technology. However, new technologies may also allow for further advances in scientific understanding. This is largely because new technologies allow us to observe the natural world in ways not previously possible. Additionally, new technology may allow us to store and analyze data in new ways. Finally, technological advances pique the interest of funding bodies and tax-payers thereby fueling further scientific investigations. Below are some examples of this phenomenon within the study of physics.

Astrophysics-space exploration
Though telescopes have been used by astronomers for hundreds of years, they have become increasingly powerful. With each improvement, scientists are able to obtain more detail about the behavior and composition of distant bodies. Dramatic advances in our knowledge of the galaxy were made during the 20th century's "space race." New developments in communications and materials allowed the creation of satellites, space probes, and manned missions that provided never before seen images of the earth and other planets and allowed new experimentation. The space race also provides us with an excellent example of how the public's interest in technology can help stimulate further scientific advances.

Biophysics-micro- and nano-instrumentation
At the other end of the size scale from telescopes, microscopes have also become more powerful over time. These have allowed us to visualize the make-up of sub-cellular structures. Even more interestingly, new tools for micro-manipulation (for example, laser tweezers) allow scientists to observe, modify and measure the behavior of proteins and nucleic acids.

Computational capabilities
Advances in computational ability have made possible the storage and manipulation of large amounts of data. These systems are then able to analyze the data for trends and correlations more rapidly and accurately than humans could. The result is highly accurate models of natural systems. Because these models are extremely complex with many variables, it is only possible to produce or use them with the assistance of powerful computers. Not only can these models be used to explain the effects of various factors, they can be used to perform simulations. In many cases, simulations allow us to investigate phenomena in which experiments are not feasible. Here again we see how the demand for new technologies can propel basic scientific research: the desire for smaller, faster electronic devices has stimulated a large amount of research into semiconductors.

Skill 1.4 **Use a variety of software and information technologies to model and solve problems in mathematics, science, and technology**

Advances in information technology, including software, have been immensely helpful in all the physical sciences. Information technology has allowed for the rapid communication of experimental results and hypotheses between scientists and engineers working in distant locations. For example, the ability to communicate rapidly by email and send digital images and other files means scientists can easily confer on new data and interpretations. They can easily find and consult with experts outside their own field who might have particular knowledge. Information technology has also made it easier to find already published experimental results; many journals have online editions and search engines make finding information on a given topic straightforward.

Specific software packages can also assist the advancement of science. Some important types of software and their appropriate uses are detailed below:

Spreadsheets

Spreadsheets contain multiple pages of grids into which numbers and formulas can be entered. They are excellent tools for storing data and doing some simple mathematical manipulations. Many spreadsheets now include graphical and statistical packages (see below), though often they are not as powerful as "stand alone" software for those uses. Spreadsheets are not typically used for complex statistical analysis or simulations though they may have some rudimentary capabilities. The data would be imported it to another, more specific program if further analysis were needed.

Databases

Like spreadsheets, databases are designed for the storage and manipulation of data. However, they are much more sophisticated and have increased security and accountability control. Though they may sometimes be useful for managing extremely large or shared datasets, databases are not typically useful to scientists.

Graphing packages

Graphing packages allow data to be transformed into a variety of visual representations. Because graphing is so important for recognizing trends and communicating experimental results, scientists make extensive use of this type of software. Most graphical packages will also perform basic modeling or statistical analysis (for instance, fitting a line or determining standard deviation). This is appropriate in some simple cases but, most often, statistical packages are better suited for this type of analysis. Graphing software should usually be limited to displaying the results of this analysis (for example, with error bars).

Statistical packages

Statistical packages perform all types of statistical analysis to determine if significant differences exist between datasets. Thus, this type of software is extremely helpful to scientists especially when one is trying to analyze trends or determine the effects of a given factor in a large, complex experiment. Though this software can simplify the performance of a statistical test, the scientist must still carefully determine *which* statistical test will provide her with the answer she seeks. Though statistical software packages may have tools for data storage and graphing, it is usually preferable to use these only temporarily or for a "first-pass" and ultimately use a more specific software package for these tasks.

Simulators

Simulators assist with the creation of complex mathematical models of the natural world. They also allow for simulated experiments in which the value of variables in the model are changed and hypothetical outcomes obtained. These simulators can be a tremendous boon to scientists when performing actual experiments would be expensive, unethical, or impossible. Though much important information can be obtained from these simulators, it is important that their findings be supported by experimental evidence whenever possible.

COMPETENCY 2.0   Understand the relationships and common themes that connect mathematics, science, and technology

Skill 2.1   Analyze the significance of key events, theories, and individuals in the history of physics

Archimedes

Archimedes was a Greek mathematician, physicist, engineer, astronomer, and philosopher. He is credited with many inventions and discoveries some of which are still in use today such as the Archimedes screw. He designed the compound pulley, a system of pulleys used to lift heavy loads such as ships.

Although Archimedes did not invent the lever, he gave the first rigorous explanation of the principles involved which are the transmission of force through a fulcrum and moving the effort applied through a greater distance than the object to be moved. His Law of the Lever states that magnitudes are in equilibrium at distances reciprocally proportional to their weights.

He also laid down the laws of flotation and described Archimedes' principle which states that a body immersed in a fluid experiences a buoyant force equal to the weight of the displaced fluid.

Niels Bohr

Bohr was a Danish physicist who made fundamental contributions to understanding atomic structure and quantum mechanics. Bohr is widely considered one of the greatest physicists of the twentieth century.
Bohr's model of the atom was the first to place electrons in discrete quantized orbits around the nucleus.

Bohr also helped determine that the chemical properties of an element are largely determined by the number of electrons in the outer orbits of the atom. The idea that an electron could drop from a higher-energy orbit to a lower one emitting a photon of discrete energy originated with Bohr and became the basis for future quantum theory.

He also contributed significantly to the Copenhagen interpretation of quantum mechanics. He received the Nobel Prize for Physics for this work in 1922.

Marie Curie

Curie was as a Polish-French physicist and chemist. She was a pioneer in radioactivity and the winner of two Nobel Prizes, one in Physics and the other in Chemistry. She was also the first woman to win the Nobel Prize.

Curie studied radioactive materials, particularly pitchblende, the ore from which uranium was extracted. The ore was more radioactive than the uranium extracted from it which led the Curies (Marie and her husband Pierre) to discover a substance far more radioactive then uranium. Over several years of laboratory work the Curies eventually isolated and identified two new radioactive chemical elements, polonium and radium. Curie refined the radium isolation process and continued intensive study of the nature of radioactivity.

Albert Einstein

Einstein was a German-born theoretical physicist who is widely considered one of the greatest physicists of all time. While best known for the theory of relativity, and specifically mass-energy equivalence, $E = mc^2$, he was awarded the 1921 Nobel Prize in Physics for his explanation of the photoelectric effect and "for his services to Theoretical Physics". In his paper on the photoelectric effect, Einstein extended Planck's hypothesis ($E = h\nu$) of discrete energy elements to his own hypothesis that electromagnetic energy is absorbed or emitted by matter in quanta and proposed a new law $E_{max} = h\nu - P$ to account for the photoelectric effect.

He was known for many scientific investigations including the special theory of relativity which stemmed from an attempt to reconcile the laws of mechanics with the laws of the electromagnetic field. His general theory of relativity considered all observers to be equivalent, not only those moving at a uniform speed. In general relativity, gravity is no longer a force, as it is in Newton's law of gravity, but is a consequence of the curvature of space-time.

Other areas of physics in which Einstein made significant contributions, achievements or breakthroughs include relativistic cosmology, capillary action, critical opalescence, classical problems of statistical mechanics and problems in which they were merged with quantum theory (leading to an explanation of the Brownian movement of molecules), atomic transition probabilities, the quantum theory of a monatomic gas, the concept of the photon, the theory of radiation (including stimulated emission), and the geometrization of physics.

Einstein's research efforts after developing the theory of general relativity consisted primarily of attempts to generalize his theory of gravitation in order to unify and simplify the fundamental laws of physics, particularly gravitation and electromagnetism, which he referred to as the Unified Field Theory.

Michael Faraday

Faraday was an English chemist and physicist who contributed significantly to the fields of electromagnetism and electrochemistry. He established that magnetism could affect rays of light and that the two phenomena were linked. It was largely due to his efforts that electricity became viable for use in technology. The unit for capacitance, the farad, is named after him as is the Faraday constant, the charge on a mole of electrons (about 96,485 coulombs). Faraday's law of induction states that a magnetic field changing in time creates a proportional electromotive force.

J. Robert Oppenheimer

Oppenheimer was an American physicist best known for his role as the scientific director of the Manhattan Project, the effort to develop the first nuclear weapons. Sometimes called "the father of the atomic bomb", Oppenheimer later lamented the use of atomic weapons. He became a chief advisor to the United States Atomic Energy Commission and lobbied for international control of atomic energy. Oppenheimer was one of the founders of the American school of theoretical physics at the University of California, Berkeley. He did important research in theoretical astrophysics, nuclear physics, spectroscopy, and quantum field theory.

Sir Isaac Newton

Newton was an English physicist, mathematician, astronomer, alchemist, and natural philosopher in the late 17th and early 18th centuries. He described universal gravitation and the three laws of motion laying the groundwork for classical mechanics. He was the first to show that the motion of objects on earth and in space is governed by the same set of mechanical laws. These laws became central to the scientific revolution that took place during this period of history. Newton's three laws of motion are:

I. Every object in a state of uniform motion tends to remain in that state of motion unless an external force is applied to it.
II. The relationship between an object's mass m, its acceleration a, and the applied force F is $F = ma$.
III. For every action there is an equal and opposite reaction.

In mechanics, Newton developed the basic principles of conservation of momentum. In optics, he invented the reflecting telescope and discovered that the spectrum of colors seen when white light passes through a prism is inherent in the white light and not added by the prism as previous scientists had claimed. Newton notably argued that light is composed of particles. He also formulated an experimental law of cooling, studied the speed of sound, and proposed a theory of the origin of stars.

Erwin Schrödinger

Schrödinger was a Nobel Prize winning Austrian physicist who is best remembered for his contributions to quantum mechanics. Chief among his findings was the Schrödinger equation which describes the space and time-dependence of quantum systems. This equation predicts behavior of microscopic particles in much the same way that Newton's second law predicts the behavior of macroscopic matter. Much of solid state physics was built upon the Schrödinger (wave) equation. Schrödinger also devised the "cat in a box" thought experiment to illustrate a paradox of quantum mechanics.

Enrico Fermi

Fermi was an extremely gifted Italian physicist and Nobel Prize winner. His work focused on radioactivity, quantum theory, and statistical mechanics and he helped develop the first nuclear reactor. He also participated in the Manhattan Project after fleeing to America prior to the Second World War. Fermi's Golden Rule is an important equation in quantum mechanics and is used to calculate the transition rates of quantum systems. Much of Fermi's work in quantum mechanics (Fermi holes, Fermi levels, Fermi-Dirac statistics) ultimately led to our modern understanding of semiconductors.

James Clerk Maxwell

Maxwell was a Scottish theoretical physicist and mathematician whose work was especially important to our understanding of electromagnetism. His famous Maxwell's equations were the first to unify and simply express the laws of electricity and magnetism. Maxwell was also important in the development of the kinetic theory; the Maxwell-Boltzman distribution is a probability distribution used in statistical mechanics and to predict the behavior of gases. His work paved the way for much of the progress in physics during the 20th century (special relativity, quantum mechanics, etc) and ultimately led to all modern electrical and communications technology.

Daniel Bernoulli

Bernoulli was a Swiss mathematician who developed many equations important in physics. His work was especially focused on the behavior of fluids and the Bernoulli equation governs all steady, inviscid, incompressible flow. The resultant Bernoulli principle is especially useful in aerodynamics. The Euler-Bernoulli beam theory, which allows calculation of the loading characteristics of a beam, was of key importance during the Industrial Revolution and remains an important equation for engineers today. Bernoulli was also the first scientist to suggest the principles of the kinetic theory of gases.

Skill 2.2   Assess the societal implications of developments in physics

Nearly all advances in science and technology have some effect on society. However, some are especially important in changing the way we think about the natural world or how we lead our day to day lives. Several examples drawn from physics are briefly explored below.

Heliocentric theory

While we largely take for granted that the planets in our solar system orbit around the sun, during the 16th and 17th centuries it was a controversial notion in the western world. Most religious authorities, at least publicly, condemned the writings of Galileo and Copernicus which postulated a heliocentric model of the solar system. Both scientists and theologians argued over how new scientific evidence could be brought into agreement with scripture. This provides a famous historical example of how organized religion can interact with, and at times impede, scientific progress. Today, we still see this playing out in debates over such theories as the Big Bang and evolution.

Nuclear physics and the atomic bomb

Advances in nuclear physics in the first half of the 20th century led to the realization that tremendous amounts of power could be generated for fuel or weaponry through atomic fission and fusion. During World War II, the United State and its allies (through the Manhattan Project) were the first to succeed in the creation of atomic bombs which were subsequently dropped on Hiroshima and Nagasaki in 1945. The long term and exceedingly devastating effect of these weapons cannot be overstated. Nuclear proliferation (especially by the US and the USSR) in the following 50 years contributed to the Cold War and created an imminent threat of vast worldwide destruction. While the Cold War has since ended, the continued testing and possession of nuclear weaponry by an ever growing number of countries remains an important socio-political problem.

Nuclear power is currently use but remains controversial. On one hand, the extreme short and long term dangers of nuclear accidents have been demonstrated by incidents such as those at Three Mile Island and Chernobyl. On the other hand, society is highly in need of a cleaner, more sustainable source of energy and nuclear power might be able to fill this role.

MEMS and nanotechnology

Many people speculate that nanotechnology is poised to make significant changes in our lives. Microelectromechanical Systems (MEMS), or micromachines are made possible by modern materials and technologies (such as lasers and micro-etching) that allow manufacture and manipulation of devices on the micrometer scale. Developments in these technologies are already being used in the automotive, biomedical, chemical, cosmetic, telecommunications, and manufacturing industries.

Semiconductors, the transistor, and digital devices

The discovery that the electrical conductive properties of semiconductors can be either permanently or temporarily manipulated has been an important one. Most importantly, it led to the creation of solid state electrical devices such as transistors. In electrical circuits, these elements replaced the previously used vacuum tubes and mechanical relays which were highly susceptible to wear and physical stressors. Transistors, however, are small, fast, accurate, reliable, and can be produced quickly and cheaply. Thus, transistors have paved for the way for the vast array of affordable electrical devices. They have been particularly important in advancing computers and digital technology. We are well aware of how digital devices and information technology, made possible in part by transistors, touch every part of our modern lives.

COMPETENCY 3.0 **Understand the process of scientific inquiry and the role of observation and experimentation in explaining natural phenomena**

Skill 3.1 **Analyze processes by which new scientific knowledge and hypotheses are generated**

The process of science always begins with a question about how the natural world works. Typically, this occurs when a scientists makes an observation and asks why something is happening, what factors affect it, or how the outcome might be different in a slightly different situation. The scientist then formulates a possible explanation for what she has observed and develops experiments to determine if her explanation is correct.

This process has been codified into a system known as the scientific method. Though application of the scientific method may vary somewhat among different scientists and in different fields, the basic steps are diagrammed below:

Observations

Scientific questions result from observations of events in nature or events observed in the laboratory. An observation is not just a look at what happens. It also includes measurements and careful records of the event. Records could include photos, drawings, or written descriptions. The observations and data collection lead to a question. In physics, observations almost always deal with the behavior of matter. Having arrived at a question, a scientist usually researches the scientific literature to see what is known about the question. Maybe the question has already been answered. The scientist then may want to test the answer found in the literature. Or, maybe the research will lead to a new question.

Hypothesis

If the question has not been answered, the scientist may prepare for an experiment by making a hypothesis. A hypothesis is a statement of a possible answer to the question. It is a tentative explanation for a set of facts and can be tested by experiments. Although hypotheses are usually based on observations, they may also be based on a sudden idea or intuition.

Experiment

An experiment tests the hypothesis to determine whether it may be a correct answer to the question or a solution to the problem. Some experiments may test the effect of one thing on another under controlled conditions. Such experiments have two variables. The experimenter controls one variable, called the independent variable. The other variable, the dependent variable, is the change caused by changing the independent variable. For example, suppose a researcher wanted to test the effect of vitamin A on the ability of rats to see in dim light.

The independent variable would be the dose of Vitamin A added to the rats' diet. The dependent variable would be the intensity of light that causes the rats to react. All other factors, such as time, temperature, age, water given to the rats, the other nutrients given to the rats, and similar factors, are held constant. Scientists sometimes do short experiments "just to see what happens". Often, these are not formal experiments. Rather they are ways of making additional observations about the behavior of matter.

In most experiments, scientists collect quantitative data, which is data that can be measured with instruments. They also collect qualitative data, descriptive information from observations other than measurements. Interpreting data and analyzing observations are important. If data is not organized in a logical manner, wrong conclusions can be drawn. Also, other scientists may not be able to follow your work or repeat your results.

Conclusion

Finally, a scientist must draw conclusions from the experiment. A conclusion must address the hypothesis on which the experiment was based. The conclusion states whether or not the data supports the hypothesis. If it does not, the conclusion should state what the experiment did show. If the hypothesis is not supported, the scientist uses the observations from the experiment to make a new or revised hypothesis. Then, new experiments are planned.

Steps of a Scientific Method

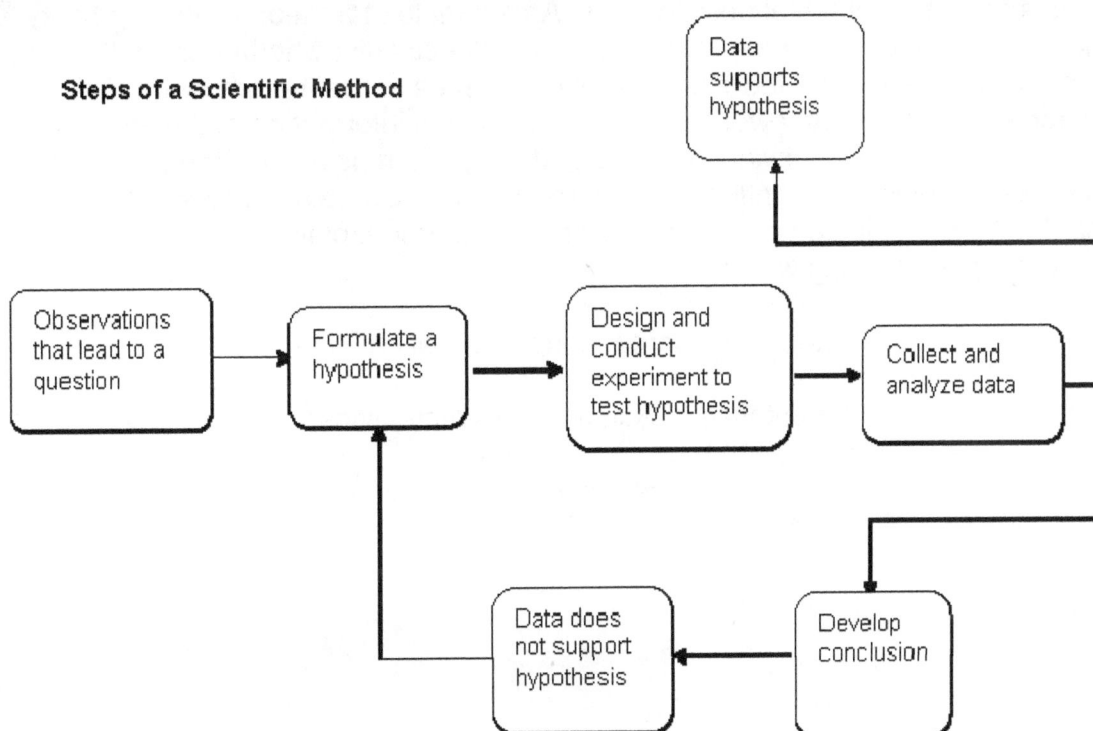

Data supports hypothesis

Observations that lead to a question → Formulate a hypothesis → Design and conduct experiment to test hypothesis → Collect and analyze data

Data does not support hypothesis ← Develop conclusion

Outcomes of the scientific method
Science is an iterative process in which hypotheses are developed and tested repeatedly. As new experimental results are obtained, old conclusions may need to be modified to fit new results. After significant experimentation, hypotheses with sufficient support may be elevated to the status of law or theory.

Law
A law is a statement of an order or relation of phenomena that, as far as is known, is invariable under the given conditions. Everything we observe in the universe operates according to known natural laws. If the truth of a statement is verified repeatedly in a reproducible way then it can reach the level of a natural law. Some well known and accepted natural laws of science are listed below:

        1.   The First Law of Thermodynamics

        2.   The Law of Cause and Effect

        3.   The Law of Biogenesis

Theory
In contrast to a law, a scientific theory is used to explain an observation or a set of observations. It is generally accepted to be true, though it cannot be proved absolutely. The important thing about a scientific theory is that there are no experimental observations to prove it NOT true, and each piece of evidence that exists supports the theory as written. Theories are often accepted at face value since they are often difficult to prove and can be rewritten in order to include the results of all experimental observations. An example of a theory is the big bang theory. While there is no experiment that can directly test whether or not the big bang actually occurred, there is no strong evidence indicating otherwise. Theories provide a framework to explain the known information of the time, but are subject to constant evaluation and updating. There is always the possibility that new evidence will conflict with a current theory. Some examples of theories that have been rejected because they are now better explained by current knowledge are list below:

        1.   Theory of Spontaneous Generation

        2.   Inheritance of Acquired Characteristics

        3.   The Blending Hypothesis

Some examples of theories that were initially rejected because they fell outside of the accepted knowledge of the time but are well-accepted today due to increased knowledge and data include:

1. The sun-centered solar system

2. The germ theory of disease

3. Continental drift

Skill 3.2   Analyze ethical issues related to the process of scientific research

Ethical behavior is critical at all points of scientific experimentation, interpretation, and communication. Both other scientists and the public in general need to be confident in the validity of scientific results and the fact that science is being performed in the service of public good.

The first aspect of this is that **science is actually performed responsibly**. It is always important that careful documents be kept of all experimental conditions and outcomes. In both commercial and academic research environments, scientists keep careful records in laboratory notebooks. These serve as primary documents bearing witness to new discoveries and developments. Additionally, it is of utmost importance that, when used, animal subjects are humanely treated and human subjects are fully aware of all relevant facts (informed consent). Most research and funding organizations have policies in place and review boards to monitor the use of animal and human subjects.

Secondly, the **scientific findings must be accurately presented**. This means that all relevant data is presented to clearly reflect experimental findings. Occasionally, a scientist will report only that data which agrees with the hypothesis he is putting forth. Alternatively, the data may be improperly statistically manipulated. These and other misrepresentations of experimental findings seriously impede the progress of science because they may provide misinformation about how the natural world operates.

Another aspect of the requirement to accurately present research is the need for the **credit for work to be properly assigned**. Whether in the development of technology or in pure research, there are both financial and scientific issues at stake. In academic and pure research settings findings are typically published in peer reviewed journals and student theses. Occasionally, cases of plagiarism or even the theft of experimental work occur. Plagiarism means that the work has been presented as original when, in fact, it incorporates the work of others. Both these occurrences are highly undesirable and may results in disciplinary action and/or loss of reputation within the scientific community. It should be noted that, when preparing documents, it is acceptable to include information and even direction quotations from other sources as long as they are always *properly cited*.

One final area in which ethics is extremely important is in **deciding in what direction scientific research should proceed**. Both scientists and the general public must determine if there are any possible investigations that *should* not actually be performed. Currently, this is especially important in medical research where appropriateness of research into human cloning, embryonic stems cells, and other topics are heavily debated.

Skill 3.3   Evaluate the appropriateness of a specified experimental design to test a given hypothesis

Disturbance by observation

Experiments designed to test physics hypotheses are plagued by a number of difficulties that can be partially, but never totally, alleviated. For example, as has been pointed out by many, an experiment does not involve a scientist making an observation of a phenomenon; rather, it involves a scientist making an observation of a phenomenon being observed by a scientist. That is to say, the act of observation has an effect on the system being observed and this must be recognized in analyzing and reporting results. This effect can be profound in microscopic experiments, especially at the level of subatomic particles where using light to observe a particle actually disturbs the particle substantially. A macroscopic example is the measurement of voltage or current in an electrical circuit. Although most voltmeters and ammeters do not present a large load to the circuit, they do load the circuit to some degree affecting the measurements. Thus, it is critical to note the inherent limitations present in an experiment when evaluating its ability to properly test some hypothesis.

Isolation of parameters

Along similar lines to the above-mentioned difficulty, it is not possible to completely isolate a particular parameter for experimental testing. Thus, for example, although theory can treat a mass as the sole entity in the universe, in reality there are virtually innumerable other masses distributed throughout the universe that have an effect (by gravitation, for example) on the mass of interest. An appropriate experiment is thus designed to minimize outside and unwanted influences to the extent possible. One should evaluate how well the experiment tests only those parameters that are of concern to the hypothesis and how well the experiment reduces, or holds invariable, other parameters that could otherwise affect the results.

Measurement precision

Another concern regarding experimental design is the ability of the equipment to measure the desired parameters with sufficient accuracy and precision. If the parameter that is being measured is beyond the range of the equipment by way of being either too large, too small, or having variations that are too fine, the experiment can yield no useful information. The measurement equipment may be more or less technical; it may simply involve the use of the human eye or it may involve sophisticated equipment such as radiation or thermal energy detectors, pressure sensors or other devices.

Financial burden

Financial concerns are no less important than physical or scientific concerns. An experiment is only as useful as the possibility of its implementation and the cost of an experiment is a critical consideration in its appropriate design. High-energy physics, for example, is hampered by this difficulty as the costs of constructing particle accelerators and other associated equipment, not to mention the costs of maintenance and personnel, can be staggeringly high. Thus, an ideal experimental design is one that can be implemented at an affordable level of expense that allows for repeated use and proper maintenance.

Skill 3.4   Assess the role of communication among scientists in promoting scientific progress

There are many important reasons for scientists to report their detailed findings to the scientific community. The first and most important reason is to **ensure the correctness of scientific results and to advance our larger understanding of the natural world**. When new scientific findings are reported at conferences or peer reviewed for publication in technical journals, they are rigorously evaluated. This is to ensure that the experiments were performed with proper controls and that the results are repeatable. In controversial situations, experiments may even be fully repeated by other scientists to ensure that the same results are obtained. Once it is established that new findings are sound, the scientists can work together to determine how these results agree or disagree with previous experiments. They may also decide together what additional investigations would be useful to the field. **When open-minded scientific discourse is supported, opinions and information can be exchanged and theories can be refined.** Scientists will weigh the experimental evidence to determine what hypothesis has the most support. Given time, new theories may emerge and scientific knowledge can both broaden and deepen.

Science has increasingly become a collaborative and interdisciplinary effort. Engineers, clinicians, and scientists may not be familiar with advances in far reaching fields unless new developments are announced to a broad community of scientists. When this happens, it opens the doors for an exchange of information that can be enormously helpful in both investigating and solving problems in science and technology. Consider, for instance, an exchange of information between physicists and biologists. The physicists may suggest the use of high powered microscopes and lasers for study of sub-cellular structures. Likewise, the biologist's knowledge of transport proteins could inform the physicist's design of new nano-machines.

It is also important that the public in general be kept informed of scientific progress. This is of key importance in any democratic society because all people together must decide the direction of the government. **New scientific discoveries can have huge implications for medical, social, and environmental issues which should be of interest to all voters**. Additionally, when people are made familiar with the advances made in science, they may be more willing to see the need for public funding which supports continued scientific progress.

COMPETENCY 4.0   Understand principles of measurement and the process of gathering, organizing, reporting, and interpreting scientific data

Skill 4.1   Evaluate the appropriateness of units of measurement, measuring devices, or methods of measurements for a given situation

In accordance with the popular saying, "the right tool for the right job," experiments require the proper equipment and approaches for measurement to be successful. A number of factors ranging from what is being measured to the context in which measurements are being taken affect how an experiment is to be conducted. Particular fields or industries, even, may have specific conventions for measurements.

SI Units

The International System of Units, or SI units, is the most common system of units used in the physical sciences. The abbreviation SI originates from the French name, *Le Système International d'Unités*. SI units are commonly accepted among most scientists for collecting and reporting measurements. In a few cases there are some exceptions with regard to SI unit usage. For instance, in the heating and air conditioning industries, the British thermal unit (BTU) is used in place of the Joule, the SI unit for energy. A system of units is always based on certain conventions, such as the precise definition of a kilogram, a meter or a second, and many other units, such as the Newton (unit of force) or the Joule (unit of energy) are derived from these definitions. No particular system of units is inherently better than another although some may be more convenient to use. SI is often a preferred system due to its ubiquity among scientists, its largely base-10 systematization and its popular use by most countries around the world. Some scientists still use the centimeter-gram-second (CGS) system of units due to its usefulness in simplifying some equations and in reducing some physical constants to unity.

Measurement devices and methods

The approach to taking measurements for a particular experiment is determined by the particular phenomenon being measured, the context of measurement and the financial limitations imposed upon the experiment. Obviously, thermometers cannot be used for measuring distances, and the right equipment must be used for the appropriate measurement. Nevertheless, more subtle considerations abound, including the required accuracy and precision of measurements. If relatively coarse measurements are sufficient, there is often no need for sophisticated, expensive equipment.

For example, if a rough measurement of weight is required, a mechanical balance scale may be all that is required. Measurements of higher precision using an advanced digital scale may be gratuitous, especially in light of other inexact measurements that may be taken during the experiment.

The particular method used for measurement is determined in large part by the theory or hypothesis being tested. Measurements in the realm of particle physics, where hypotheses are largely based on a synthesis of quantum mechanics and special relativity, do not allow for the use of approaches to measurement that assume classical mechanics and electrodynamics. Instead, the particular method used must be based on more firmly established principles and concepts from the theory. Additionally, certain types of measurements must be approached indirectly. That is to say, empirical information about some parameters of a system may actually require measurement of different parameters from which measurements the desired parameter can be calculated. Such indirect measurements may be required for phenomena that are newly discovered, that are microscopic in scale or that require extremely high-precision results.

Skill 4.2   Assess the appropriateness of a given method or procedure for collecting data for a specified purpose

Resistance to human error
A number of factors influence the appropriateness of data collection methods with regard to the purpose for which they are being employed. The likelihood of human error must be screened carefully, as mistakes in setting up equipment, transcription errors in recording data or disturbances of the setup during data collection can all lead to skewed results. In some situations, it may be sufficient for the scientist to simply take measurements and record them by hand, but, at other times, a computer or data acquisition system may be required. In many electromagnetics applications, for example, the data rates or frequencies of operation are often so high as to make automated data collection a necessity.

Sufficient sampling
The data collection method must be able sample data at a sufficiently high rate to adequately characterize the process or phenomenon being studied. In the case of transient wave phenomena, such as transient electromagnetic or acoustic signals, the data collection rate must be at least twice as high as the highest frequency component of the signal. This so-called Nyquist rate is the minimum threshold for preventing aliasing in the collected data. The minimum requirements for data sampling may be temporal, as with many acoustic and electromagnetic signals, or they may be spatial, as with remote sensing, for example.

Sufficient data storage

Along similar lines, sufficient storage space for data must be available. Often, this involves computer storage in the form of a hard drive, or, perhaps, an optical medium. Data acquisition systems operating at high rates can produce tremendous amounts of data in short periods of time. Financial and even space considerations apply when a large number of CDs, DVDs or other media are required for storage.

Data precision

The precision of the data collection approach must also be evaluated. Both the sensing or measurement device, as well as the digitization equipment (if computer-based data acquisition is being used), must be of sufficient precision to measure the parameters of the system of interest to accurately record variations in those parameters. If, for example, a transducer is too coarse in its measurement abilities or an analog-to-digital converter does not have a sufficient number of bits to accurately digitize the signal, then information is lost. Similarly, the data that is collected should not be assigned more precision than the equipment allows. If the digitization equipment is more precise than the measurement or transduction equipment, for example, then the data collected may have more numerical precision than the transducer could have possibly afforded.

Skill 4.3 Assess the use of statistical methods for analyzing data

Measurements of data are subject to numerous types of errors (or "noise") that can lead to variations in results that might otherwise not have been expected. These **variations in measurement data can result from inherent noise in equipment, human error or other factors**. Since no one measurement or datum can generally be treated as any more correct than another measurement or datum, as the variations are often random, the aggregate data must be treated statistically.

The effect of the use of statistics in analyzing data is to remove or quantify factors that adversely affect the precision of a set of measurements or other data. Alternatively, statistics can be used to provide a more compact description of a set of data corresponding to random variables.

An example experiment is the rolling of a ball on a flat surface: the ball is launched with a certain force and the distance traveled is measured to gain some insight into friction. The ball will likely travel a different distance for each trial, although the variation may be small. These differences can result from an inability of the launching device (perhaps the human hand) to provide the same impulse for each instance, from small, random variations in the texture of the surface, or from measurement error during the determination of the distance traveled.

Analysis of the aggregate data by calculating a mean or median value and the variance or standard deviation of the measurements, for example, can provide a more informative assessment of the results than can any particular measurement datum.

For a set of n data values $x_1, x_2, ... x_n$, some of the common statistical parameters calculated are:

Mean given by $\bar{x} = \dfrac{\sum x_i}{n}$

Median or middle value in an ordered list of data
Mode or most frequently occurring value

Standard deviation given by $\sqrt{\dfrac{\sum (x_i - \bar{x})^2}{n}}$

For sufficiently large measurement (or other data) populations, **statistical distributions** can be created to make probabilistic predictions about future measurement results or responses of a system. This is crucial for such situations as, for example, probability of failure for some system. The accuracy or precision of a measurement device can also be treated statistically.

Proper statistical analysis of data is crucial, especially when that data is used to calculate other parameters. Error propagation (or uncertainty propagation) must be treated carefully, as the variation in a measured parameter, such as the position of a moving car, cannot be directly translated into the variation of a calculated parameter, such as the acceleration of that car. The error, or uncertainty, of the acceleration must be appropriately calculated based on the error of the position measurements; the two errors are generally not the same value.

Statistical analysis can be used for complex systems when exact treatment of each individual aspect of the system is impossible. **A statistical analysis allows for the determination of the behavior of a system on average, thus providing some insight when an exact analysis is too costly in terms of time or other resources**.

Skill 4.4 Analyze relationships among factors as indicated by experimental data

In many experimental investigations we attempt to understand the relationship between two variables. Typically, one variable (independent, usually represented x) is altered while the other is measured (dependent, usually represented y). Certain relationships between these variables are commonly observed. Often, the relationships are most clearly observable when graphed.

Linear
In linear relationships, the independent variable is *directly* proportional to the dependent variable. This means that, as the absolute value of x is increased, the absolute value of y will also increase. This is stated mathematically as:

$$y \propto x$$

This relationship can also be represented using the familiar equation for a straight line (where m is slope and b is the intercept):

$$y = mx + b$$

Graphs of this type of relationship are shown below. On the left, the slope of the line (m) is positive and on the right it is negative:

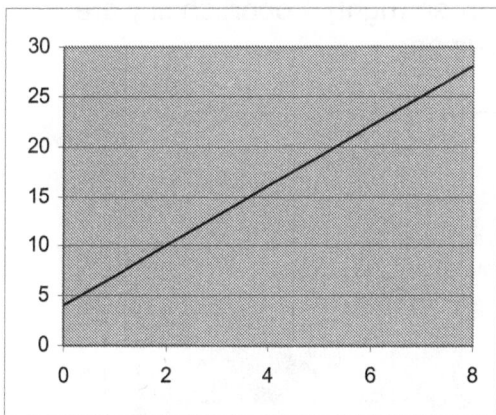

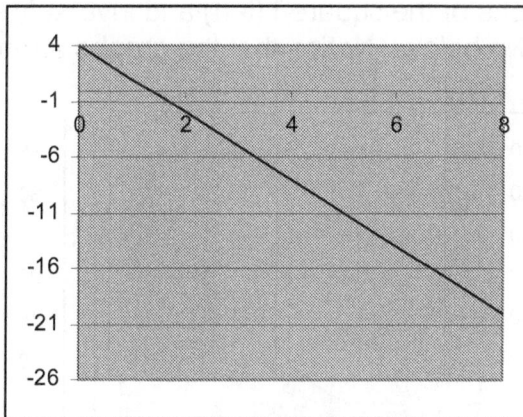

Inverse
Unlike direct proportionality, inverse proportionality signifies that the dependent variable is a function of the inverse of the independent variable:

$$y \propto \frac{1}{x}$$

A graph of this relationship is hyperbolic, as shown below:

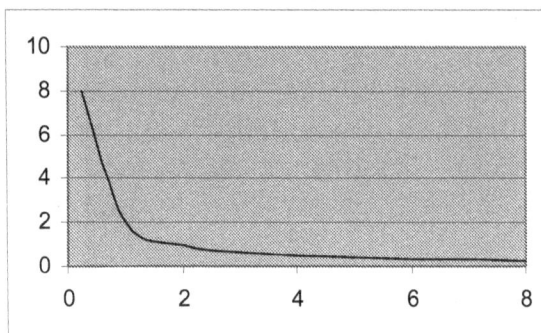

Squared and Inverse Squared
As the name suggests, in squared relationships the dependent variable is proportional to the square of the independent variable:

$$y \propto x^2$$

Similarly to the inverse proportionality described above, the dependent variable may also be proportional to the inverse square of the independent variable:

$$y \propto \frac{1}{x^2}$$

Graphs of the squared (left) and inversely squared (right) proportionality are shown below. Notice that the graphs are non-linear.

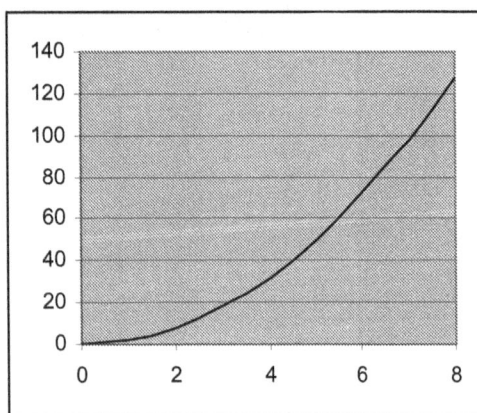

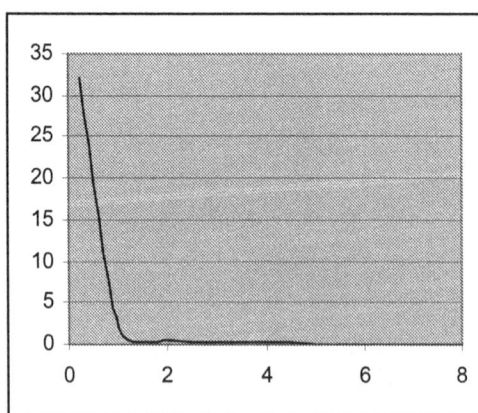

Logarithmic and Exponential

The dependent variable may also be proportional to a logarithmic or exponential function of the independent variable.

$$y \propto \log_a x \qquad y \propto a^x$$

Graphs of these relationships are shown below (logarithmic relationship on the left and exponential on the right). Make special note of the scale of the dependent variable.

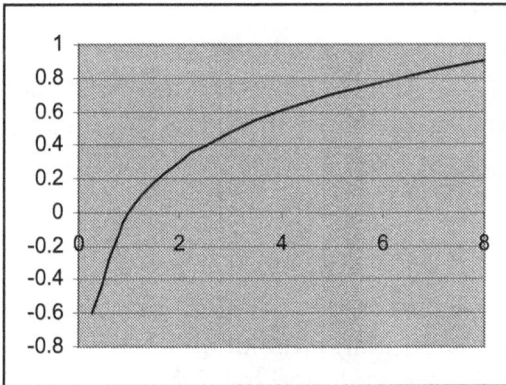

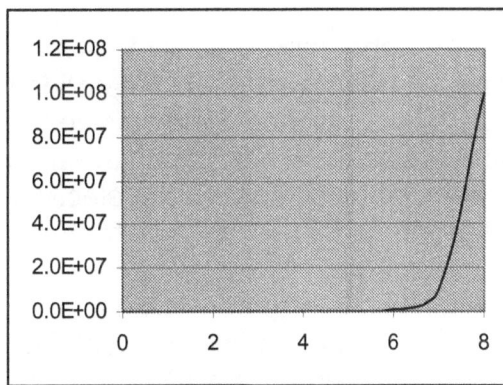

COMPETENCY 5.0   Understand equipment, materials, and chemicals used in science investigations; and apply procedures for their proper and safe use

Skill 5.1   Analyze the principles upon which given laboratory instruments are based

Oscilloscope: An oscilloscope is a piece of electrical test equipment that allows signal voltages to be viewed as two-dimensional graphs of electrical potential differences plotted as a function of time.

The oscilloscope functions by measuring the deflection of a beam of electrons traveling through a vacuum in a cathode ray tube. The deflection of the beam can be caused by a magnetic field outside the tube or by electrostatic energy created by plates inside the tube. The unknown voltage or potential energy difference can be determined by comparing the electron deflection it causes to the electron deflection caused by a known voltage.

Oscilloscopes can also determine if an electrical circuit is oscillating and at what frequency. They are particularly useful for troubleshooting malfunctioning equipment. You can see the "moving parts" of the circuit and tell if the signal is being distorted. With the aid of an oscilloscope you can also calculate the "noise" within a signal and see if the "noise" changes over time.

Inputs of the electrical signal are usually entered into the oscilloscope via a coaxial cable or probes. A variety of transducers can be used with an oscilloscope that enable it to measure other stimuli including sound, pressure, heat, and light.

Signal Generator: A signal generator, also known as a test signal generator, function generator, tone generator, arbitrary waveform generator, or frequency generator, is a device that generates repeating electronic signals in either the analog or digital domains. They are generally used in designing, testing, troubleshooting, and repairing electronic devices.

A function generator produces simple repetitive waveforms by utilizing a circuit called an electronic oscillator or a digital signal processor to synthesize a waveform. Common waveforms are sine, sawtooth, step or pulse, square, and triangular. Arbitrary waveform generators are also available which allow a user to create waveforms of any type within the frequency, accuracy and output limits of the generator. Function generators are typically used in simple electronics repair and design where they are used to stimulate a circuit under test. A device such as an oscilloscope is then used to measure the circuit's output.

Spectrometer: A spectrometer is an optical instrument used to measure properties of light over a portion of the electromagnetic spectrum. Light intensity is the variable that is most commonly measured but wavelength and polarization state can also be determined. A spectrometer is used in spectroscopy for producing spectral lines and measuring their wavelengths and intensities. Spectrometers are capable of operating over a wide range of wavelengths, from short wave gamma and X-rays into the far infrared. In optics, a spectrograph separates incoming light according to its wavelength and records the resulting spectrum in some detector. In astronomy, spectrographs are widely used with telescopes.

Geiger counter: A Geiger counter detects ionizing radiation, i.e. radiation capable of ionizing atoms and molecules and causing damage to DNA and living tissue. The core of a Geiger counter is a Geiger-Müller tube, a tube filled with inert gas (typically helium, neon or argon) and containing a thin metal wire carrying a charge. When a nuclear particle penetrates the tube, it temporarily makes the gas conductive and electrons are ultimately attracted to the central conducting wire. As the particle travels towards the wire, it continues to collide with atoms and release more electrons, thereby amplifying the signal. Ultimately, a pulse is produced by each particle that enters the detector. The counters are designed to display this signal with a needle or lamp or to produce audible clicks.

The earliest device of this type was produced by Ernest Rutherford in 1908 and the design was later refined by Hans Geiger and Walther Müller 20 years later. They are presently called "counters" because every particle produces a pulse or click, allowing the total number of particles to be tabulated. Geiger counters are commonly used because they are sturdy and relatively cheap to produce. However, they are only capable of determining the intensity of radiation (number of particles present over a given time). The energy level of the particles is also an important quantity but basic Geiger counters are not able to measure this.

Skill 5.2   Analyze hazards associated with given laboratory materials

Motion and forces: All stationary devices must be secured by C-clamps. Protective goggles must be used. Care must be taken at all times while knives, glass rods and heavy weights are used. Viewing a solar eclipse must always be indirect. When using model rockets, NASA's safety code must be implemented.

Heat: The master gas valve must be off at all times except while in use. Goggles and insulated gloves are to be used whenever needed. Never use closed containers for heating. Burners and gas connections must be checked periodically. Gas jets must be closed soon after the experiment is over. Fire retardant pads and quality glassware such as Pyrex must be used.

Pressure: While using a pressure cooker, never allow pressure to exceed 20 lb/square inch. The pressure cooker must be cooled before it is opened. Care must be taken when using mercury since it is poisonous. A drop of oil on mercury will prevent the mercury vapors from escaping.

Light: Broken mirrors or those with jagged edges must be discarded immediately. Sharp-edged mirrors must be taped. Spectroscopic light voltage connections must be checked periodically. Care must be taken while using ultraviolet light sources. Some students may have psychological or physiological reactions to the effects of strobe like (e.g. epilepsy).

Lasers: Direct exposure to lasers must not be permitted. The laser target must be made of non-reflecting material. The movement of students must be restricted during experiments with lasers. A number of precautions while using lasers must be taken – use of low power lasers, use of approved laser goggles, maintaining the room's brightness so that the pupils of the eyes remain small. Appropriate beam stops must be set up to terminate the laser beam when needed. Prisms should be set up before class to avoid unexpected reflection.

Sound: Fastening of the safety disc while using the high speed siren disc is very important. Teacher must be aware of the fact that sounds higher than 110 decibels will cause damage to hearing.

Radiation: Proper shielding must be used while doing experiments with x-rays. All tubes that are used in a physics laboratory such as vacuum tubes, heat effect tubes, magnetic or deflection tubes must be checked and used for demonstrations by the teacher. Cathode rays must be enclosed in a frame and only the teacher should move them from their storage space. Students must watch the demonstration from at least eight feet away.

Radioactivity: The teacher must be knowledgeable and properly trained to handle the equipment and to demonstrate. Proper shielding of radioactive material and proper handling of material are absolutely critical. Disposal of any radioactive material must comply with the guidelines of NRC.

Chemicals: All laboratory solutions should be prepared as directed in the lab manual. Care should be taken to avoid contamination. All glassware should be rinsed thoroughly with distilled water before using and cleaned well after use. Unused solutions should be disposed of according to local disposal procedures. Chemicals should not be stored on bench tops or heat sources. They should be stored in groups based on their reactivity with one another and in protective storage cabinets. All containers within the lab must be labeled. Chemical waste should be disposed of in properly labeled containers. Waste should be separated based on their reactivity with other chemicals. **Material safety data sheets** are available for every chemical substance directly from the company of acquisition or the internet.

The following chemicals are potential carcinogens and not allowed in school facilities: Acrylonitriel, Arsenic compounds, Asbestos, Bensidine, Benzene, Cadmium compounds, Chloroform, Chromium compounds, Ethylene oxide, Ortho-toluidine, Nickel powder, and Mercury.

It is important that teachers and educators follow these guidelines to protect the students and to avoid most of the hazards. They have a responsibility to protect themselves as well. **There should be not any compromises in issues of safety.**

Skill 5.3   Apply proper procedures for safety in the laboratory

All laboratory solutions should be prepared as directed in the lab manual. Care should be taken to avoid contamination. All glassware should be rinsed thoroughly with distilled water before using and cleaned well after use. All solutions should be made with distilled water as tap water contains dissolved particles that may affect the results of an experiment. Unused solutions should be disposed of according to local disposal procedures.

The "Right to Know Law" covers science teachers who work with potentially hazardous chemicals. Briefly, the law states that employees must be informed of potentially toxic chemicals. An inventory must be made available if requested. The inventory must contain information about the hazards and properties of the chemicals. This inventory is to be checked against the "Substance List". Training must be provided on the safe handling and interpretation of the **Material Safety Data Sheet (MSDS)**.

The following chemicals are potential carcinogens and not allowed in school facilities: Acrylonitriel, Arsenic compounds, Asbestos, Bensidine, Benzene, Cadmium compounds, Chloroform, Chromium compounds, Ethylene oxide, Ortho-toluidine, Nickel powder, and Mercury.

Chemicals should not be stored on bench tops or heat sources. They should be stored in groups based on their reactivity with one another and in protective storage cabinets. All containers within the lab must be labeled. Suspect and known carcinogens must be labeled as such and segregated within trays to contain leaks and spills.

Chemical waste should be disposed of in properly labeled containers. Waste should be separated based on their reactivity with other chemicals.

Biological material should never be stored near food or water used for human consumption. All biological material should be appropriately labeled. All blood and body fluids should be put in a well-contained container with a secure lid to prevent leaking. All biological waste should be disposed of in biological hazardous waste bags.

Material safety data sheets are available for every chemical and biological substance. These are available directly from the company of acquisition or the internet. The manuals for equipment used in the lab should be read and understood before using them.

Safety is a learned behavior and must be incorporated into instructional plans. Measures of prevention and procedures for dealing with emergencies in hazardous situations have to be in place and readily available for reference. Copies of these must be given to all people concerned, such as administrators and students.

The single most important aspect of safety is planning and anticipating various possibilities and preparing for the eventuality. Any Physics teacher/educator planning on doing an experiment must try it before the students do it. In the event of an emergency, quick action can prevent many disasters. The teacher/educator must be willing to seek help at once without any hesitation because sometimes it may not be clear that the situation is hazardous and potentially dangerous.

There are a number of procedures to prevent and correct any hazardous situation. There are several safety aids available commercially such as posters, safety contracts, safety tests, safety citations, texts on safety in secondary classroom/laboratories, hand books on safety and a host of other equipment. Another important thing is to check the laboratory and classroom for safety and report it to the administrators before staring activities/experiments. It is important that teachers and educators follow these guidelines to protect the students and to avoid most of the hazards. They have a responsibility to protect themselves as well. **There should be not any compromises in issues of safety.**

Students should wear safety goggles when performing dissections, heating, or while using acids and bases. Hair should always be tied back and objects should never be placed in the mouth. Food should not be consumed while in the laboratory. Hands should always be washed before and after laboratory experiments. In case of an accident, eye washes and showers should be used for eye contamination or a chemical spill that covers the student's body. Small chemical spills should only be contained and cleaned by the teacher. Kitty litter or a chemical spill kit should be used to clean spill. For large spills, the school administration and the local fire department should be notified. Biological spills should only be handled by the teacher. Contamination with biological waste can be cleaned by using bleach when appropriate.

Accidents and injuries should always be reported to the school administration and local health facilities. The severity of the accident or injury will determine the course of action to pursue.

All science labs should contain the following items of **safety equipment**.

- Fire blanket that is visible and accessible and signs designating room exits
- Ground Fault Circuit Interrupters (GCFI) within two feet of water supplies
- Emergency shower providing a continuous flow of water
- Emergency eye wash station that can be activated by the foot or forearm
- Eye protection for every student and a means of sanitizing equipment
- Emergency exhaust fans providing ventilation to the outside of the building
- An ABC fire extinguisher and chemical spill control kit
- Storage cabinets for flammables and fume hood with a spark proof motor
- Protective laboratory aprons made of flame retardant material
- Signs that will alert potential hazardous conditions
- Labeled containers for broken glassware, flammables, corrosives, and waste.
- Master cut-off switches for gas, electric and compressed air. Switches must have permanently attached handles. Cut-off switches must be clearly labeled.

Skill 5.4 Apply proper procedures for dealing with accidents and injuries in the laboratory

Safety is a learned behavior and must be incorporated into instructional plans. Measures of prevention and procedures for dealing with emergencies in hazardous situations have to be in place and readily available for reference. Copies of these must be given to all people concerned, such as administrators and students.

The single most important aspect of safety is planning and anticipating various possibilities and preparing for the eventuality. Any Physics teacher/educator planning on doing an experiment must try it before the students do it. In the event of an emergency, quick action can prevent many disasters. The teacher/educator must be willing to seek help at once without any hesitation because sometimes it may not be clear that the situation is hazardous and potentially dangerous. Accidents and injuries should always be reported to the school administration and local health facilities. The severity of the accident or injury will determine the course of action to pursue.

There are a number of procedures to prevent and correct any hazardous situation. There are several safety aids available commercially such as posters, safety contracts, safety tests, safety citations, texts on safety in secondary classroom/laboratories, hand books on safety and a host of other equipment. Another important thing is to check the laboratory and classroom for safety and report it to the administrators before staring activities/experiments.

SUBAREA II.   **MECHANICS AND THERMODYNAMICS**

COMPETENCY 6.0   **Understand concepts related to motion in one and two dimensions, and solve problems that require the use of algebra, calculus, and graphing**

Skill 6.1   **Apply the terminology, units, and equations used to describe and analyze one- and two-dimensional motion**

Kinematics is the part of mechanics that seeks to understand the motion of objects, particularly the relationship between position, velocity, acceleration and time.

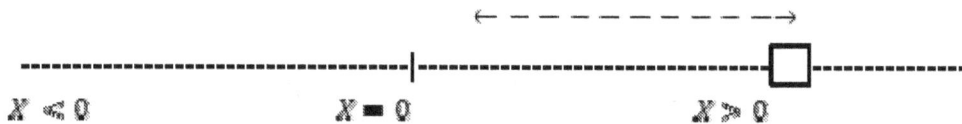

The above figure represents an object and its displacement along one linear dimension.

First we will define the relevant terms:

1. Position or Distance is usually represented by the variable x. It is measured relative to some fixed point or datum called the origin in linear units, meters, for example.

2. Displacement is defined as the change in position or distance which an object has moved and is represented by the variables D, d or Δx. Displacement is a vector with a magnitude and a direction.

3. Velocity is a vector quantity usually denoted with a V or v and defined as the rate of change of position. Typically units are distance/time, m/s for example. Since velocity is a vector, if an object changes the direction in which it is moving it changes its velocity even if the speed (the scalar quantity that is the magnitude of the velocity vector) remains unchanged.

     i) Average velocity: $\vec{v} = \frac{\Delta d}{\Delta t} = d_1 - d_0 / t_1 - t_0$.

     The ratio, $\Delta d / \Delta t$ is called the average velocity.
     Average here denotes that this quantity is defined over a period Δt.

     ii) Instantaneous velocity is the velocity of an object at a particular moment in time. Conceptually, this can be imagined as the extreme case when Δt is infinitely small.

4. Acceleration represented by *a* is defined as the rate of change of velocity and the units are m/s^2. Both an average and an instantaneous acceleration can be defined similarly to velocity.

From these definitions we develop the kinematic equations. In the following, subscript *i* denotes initial and subscript *f* denotes final values for a time period. Acceleration is assumed to be constant with time.

$$v_f = v_i + at \qquad (1)$$

$$d = v_i t + \frac{1}{2}at^2 \qquad (2)$$

$$v_f^{\,2} = v_i^{\,2} + 2ad \qquad (3)$$

$$d = \left(\frac{v_i + v_f}{2}\right)t \qquad (4)$$

In two dimensions the same relationships apply, but each dimension must be treated separately.

Skill 6.2   Analyze the motion of freely falling objects near the surface of the earth

The most common example of an object moving near the surface of the earth is a projectile. A projectile is an object upon which the only force acting is gravity. Some examples:

  i) An object dropped from rest
  ii) An object thrown vertically upwards at an angle
  iii) A canon ball

Once a projectile has been put in motion (say, by a canon or hand) the only force acting on it is gravity which near the surface of the earth is characterized by the acceleration a=g=9.8m/s^2.

This is most easily considered with an example such as the case of a bullet shot horizontally from a standard height at the same moment that a bullet is dropped from exactly the same height. Which will hit the ground first? If we assume wind resistance is negligible, then the acceleration due to gravity is our only acceleration on either bullet and we must conclude that they will hit the ground at the same time. The horizontal motion of the bullet is not affected by the downward acceleration.

Example:
I shoot a projectile at 1000 m/s from a perfectly horizontal barrel exactly 1 m above the ground. How far does it travel before hitting the ground?

Solution:
First figure out how long it takes to hit the ground by analyzing the motion in the vertical direction. In the vertical direction, the initial velocity is zero so we can rearrange kinematic equation 2 from the previous section to give:

$t = \sqrt{\dfrac{2d}{a}}$. Since our displacement is 1 m and a = g = 9.8m/s², t = 0.45 s.

Now use the time to hitting the ground from the previous calculation to calculate how far it will travel horizontally. Here the velocity is 1000m/s and there is no acceleration. So we simple multiply velocity with time to get the distance of 450m.

Skill 6.3   Solve problems involving distance, displacement, speed, velocity, and acceleration

Simple problems involving distance, displacement, speed, velocity, and constant acceleration can be solved by applying the kinematics equations from the proceeding section. The following steps should be employed to simplify a problem and apply the proper equations:

1. Create a simple diagram of the physical situation.
2. Ascribe a variable to each piece of information given.
3. List the unknown information in variable form.
4. Write down the relationships between variables in equation form.
5. Substitute known values into the equations and use algebra to solve for the unknowns.
6. Check your answer to ensure that it is reasonable.

Example:
A man in a truck is stopped at a traffic light. When the light turns green, he accelerates at a constant rate of 10 m/s². **a)** How fast is he going when he has gone 100 m? **b)** How fast is he going after 4 seconds? **c)** How far does he travel in 20 seconds?

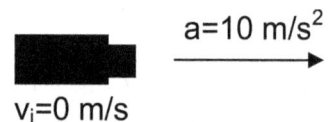

a=10 m/s²

v_i=0 m/s

Solution:
We first construct a diagram of the situation. In this example, the diagram is very simple, only showing the truck accelerating at the given rate. Next we define variables for the known quantities: a=10 m/s²; v_i=0 m/s (shown in diagram).

Now we will analyze each part of the problem, continuing with the process outlined above.

For part **a),** we have one additional known variable: d=100 m

The unknowns are: v_f (the velocity after the truck has traveled 100m)

Equation (3) will allow us to solve for v_f, using the known variables:

$$v_f^2 = v_i^2 + 2ad$$

$$v_f^2 = (0m/s)^2 + 2(10m/s^2)(100m) = 2000\frac{m^2}{s^2}$$

$$v_f = 45\frac{m}{s}$$

We use this same process to solve part **b)**. We have one additional known variable: t=4 s

The unknowns are: v_f (the velocity after the truck has traveled for 4 seconds)

Thus, we can use equation (1) to solve for v_f:

$$v_f = v_i + at$$

$$v_f = 0m/s + (10m/s^2)(4s) = 40m/s$$

For part **c)**, we have one additional known variable: t= 20 s

The unknowns are: d (the distance after the truck has traveled for 20 seconds)

Equation (2) will allow us to solve this problem:

$$d = v_i t + \frac{1}{2}at^2$$

$$d = (0m/s)(20s) + \frac{1}{2}(10m/s^2)(20s)^2 = 2000m$$

Finally, we consider whether these solutions seem physically reasonable. In this simple problem, we can easily say that they do.

Skill 6.4   Interpret information presented in one or more graphic representations related to distance, displacement, speed, velocity, and constant acceleration

The relationship between time, position or distance, velocity and acceleration can be understood conceptually by looking at a graphical representation of each as a function of time. Simply, the velocity is the slope of the position vs. time graph and the acceleration is the slope of the velocity vs. time graph. If you are familiar with calculus then you know that this relationship can be generalized: velocity is the first derivative and acceleration the second derivative of position. Here are three examples:

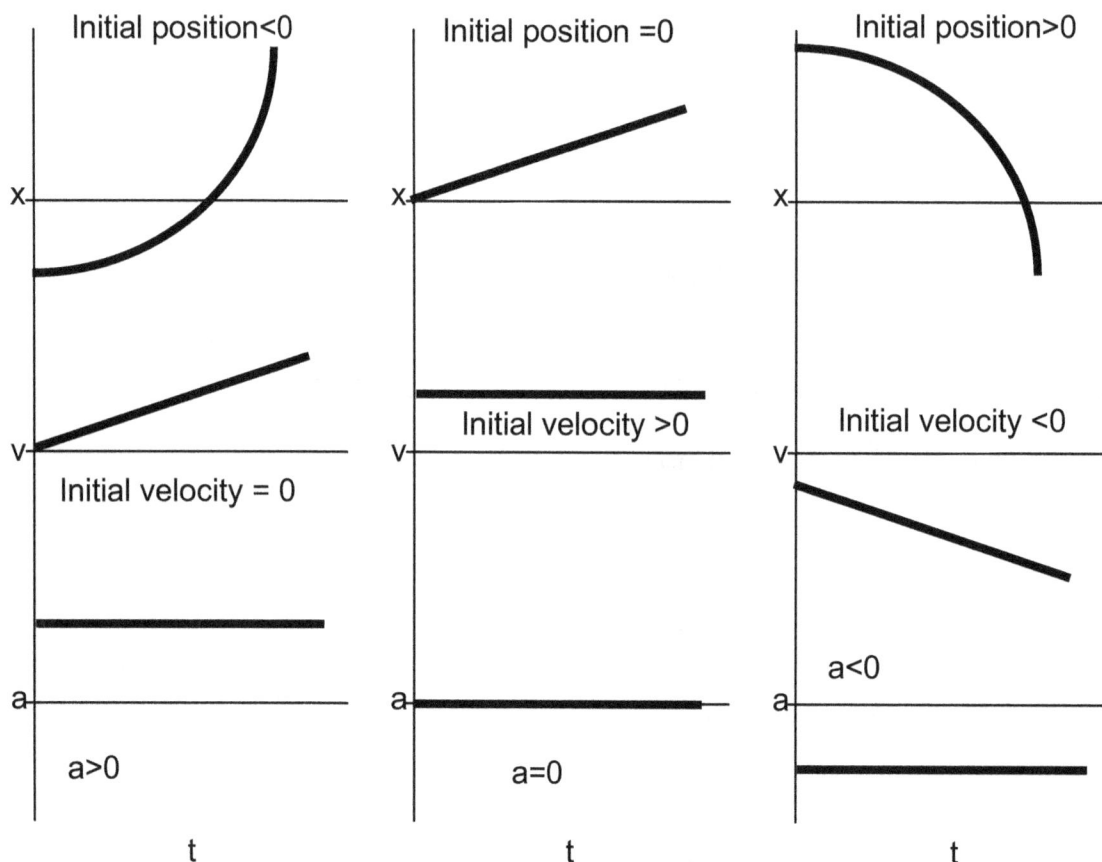

There are three things to notice:

  1) In each case acceleration is constant. This isn't always the case, but a simplification for this illustration.

  2) A non-zero acceleration produces a position curve that is a parabola.

  3) In each case the initial velocity and position are specified separately. The acceleration curve gives the shape of the velocity curve, but not the initial value and the velocity curve gives the shape of the position curve but not the initial position.

COMPETENCY 7.0 **Understand characteristics of forces and methods used to measure force, and solve problems involving forces**

Skill 7.1 **Identify the forces acting in a given situation**

Some of the common forces that act on a body are the following:

Gravity
This is the force that pulls a body towards the center of the earth, i.e. downwards, and is also called the weight of the body. It is given by

$$W = mg$$

Where m is the mass of the body and g = 9.81 m/s² is the acceleration due to gravity.

Normal force
When a body is pressed against a surface it experiences a reaction force that is perpendicular to the surface and in the direction away from the surface. For instance, an object resting on a table experiences an upward reaction force from the table that is equal and opposite to the force that the object exerts on the table. When the table is horizontal and no additional force is being applied to the object, the normal force is equal to the weight of the object.

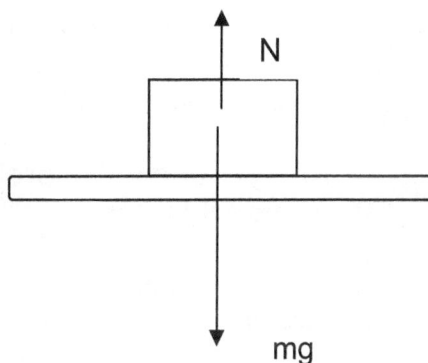

Friction
Friction is the force on a body that opposes its sliding over a surface. This force is due to the bonding between the two surfaces and is greater for rough surfaces. It acts in the direction opposite to the force attempting to move the object. When the object is at rest, the frictional force is known as **static friction**. The frictional force on an object in motion is known as **kinetic friction**.

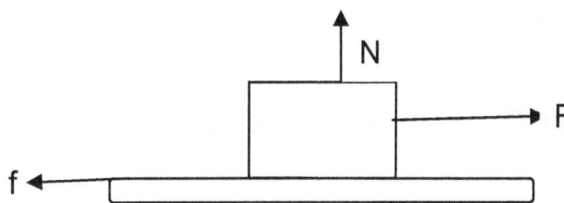

The frictional force is usually directly proportional to the normal force and can be calculated as $\mathbf{F_f} = \mu\ \mathbf{F_n}$ where μ is either the coefficient of static friction or kinetic friction depending on whether the object is at rest or in motion.

Tension/Compression

Tension is the force that acts in a rope, cable or rod that is attached to something and is being pulled. Tension acts along the cord. When a hand pulls a rope attached to a box, for instance, the tension T in the rope acts to pull the rope apart while it works on the box and the hand in the opposite direction as shown below:

Compression is the opposite of tension in that the force acts to shorten a rigid body instead of pulling it apart.

Pressure

For a description of fluid pressure and the force of buoyancy.

Net force

The forces that act on a body come from many different sources. Their effect on a body, however, is the same; a change in the state of motion of the body as given by Newton's laws of motion. Therefore, once we identify the magnitude and direction of each force acting on a body, we can combine the effect of all the forces together using vector addition and find the net force.

Skill 7.2   Analyze and solve problems involving gravitational and frictional forces

Newton's universal law of gravitation states that any two objects experience a force between them as the result of their masses. Specifically, the force between two masses m_1 and m_2 can be summarized as

$$F = G\frac{m_1 m_2}{r^2}$$

where G is the gravitational constant ($G = 6.672 \times 10^{-11} \, Nm^2 \, / \, kg^2$), and r is the distance between the two objects.

The weight of an object is the result of the gravitational force of the earth acting on its mass. The acceleration due to Earth's gravity on an object is 9.81 m/s². Since force equals mass * acceleration, the magnitude of the gravitational force created by the earth on an object is

$$F_{Gravity} = m_{object} \cdot 9.81 \, m\!\!\!/_{s^2}$$

Important things to remember:

1. The gravitational force is proportional to the masses of the two objects, but *inversely* proportional to the *square of the distance* between the two objects.

2. When calculating the effects of the acceleration due to gravity for an object above the earth's surface, the distance above the surface is ignored because it is inconsequential compared to the radius of the earth. The constant figure of 9.81 m/s^2 is used instead.

Problem: Two identical 4 kg balls are floating in space, 2 meters apart. What is the magnitude of the gravitational force they exert on each other?

Solution:

$$F = G\frac{m_1 m_2}{r^2} = G\frac{4 \times 4}{2^2} = 4G = 2.67 \times 10^{-10}\, N$$

In the real world, whenever an object moves its motion is opposed by a force known as friction. How strong the frictional force is depends on numerous factors such as the roughness of the surfaces (for two objects sliding against each other) or the viscosity of the liquid an object is moving through. Most problems involving the effect of friction on motion deal with sliding friction. This is the type of friction that makes it harder to push a box across cement than across a marble floor.

When you try and push an object from rest, you must overcome the maximum **static friction** force to get it to move. Once the object is in motion, you are working against **kinetic friction** which is smaller than the static friction force previously mentioned. Sliding friction is primarily dependent on two things, the **coefficient of friction (μ)** which is primarily dependent on roughness of the surfaces involved and the amount of force pushing the two surfaces together. This force is also known as the **normal force (F_n)**, the perpendicular force between two surfaces. When an object is resting on a flat surface, the normal force is pushing opposite to the gravitational force – straight up. When the object is resting on an incline, the normal force is less (because it is only opposing that portion of the gravitational force acting perpendicularly to the object) and its direction is perpendicular to the surface of incline but at an angle from the ground. Therefore, for an object experiencing no external action, the magnitude of the normal force is either equal to or less than the magnitude of the gravitational force (F_g) acting on it. The frictional force (F_f) acts perpendicularly to the normal force, opposing the direction of the object's motion.

The frictional force is normally directly proportional to the normal force and, unless you are told otherwise, can be calculated as $F_f = \mu F_n$ where μ is either the coefficient of static friction or kinetic friction depending on whether the object starts at rest or in motion. In the first case, the problem is often stated as "how much force does it take to start an object moving" and the frictional force is given by $F_f > \mu_s F_n$ where μ_s is the coefficient of static friction. When questions are of the form "what is the magnitude of the frictional force opposing the motion of this object," the frictional force is given by $F_f = \mu_k F_n$ where μ_k is the coefficient of kinetic friction. There are several important things to remember when solving problems about friction.

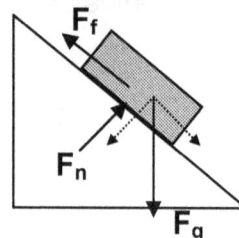

1. The frictional force acts in opposition to the direction of motion.
2. The frictional force is proportional, and acts perpendicular to, the normal force.
3. The normal force is perpendicular to the surface the object is lying on. If there is a force pushing the object against the surface, it will increase the normal force.

Problem: A woman is pushing an 800N box across the floor. She pushes with a force of 1000 N. The coefficient of kinetic friction is 0.50. If the box is already moving, what is the force of friction acting on the box?

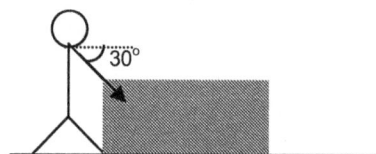

Solution: First solve for the normal force:
$F_n = 800N + 1000N (\sin 30^\circ) = 1300N$
Then, since $F_f = \mu F_n = 0.5*1300 = 650N$

Skill 7.3   Find the resultant force in a given situation

Problem:
Find the net force on a 5 Kg box sliding down an inclined surface at an angle of 30^0 with the horizontal if the coefficient of friction of the surface is 0.5.

Solution:
There are three forces acting on the box, gravity, the normal force and the frictional force. We can resolve these forces along the inclined plane and perpendicular to the plane and find the net force in each direction.

Perpendicular to the plane:
The component of the gravitational force perpendicular to the plane = mgcos30 = 5 x 9.8 x 0.87 = 42.6N

The normal force acting on the box is equal and opposite to the perpendicular component of the gravitational force. Thus the net force on the box perpendicular to the plane is zero.

Along the inclined plane:
The component of the gravitational force down the plane =
= mgsin30
= 5 x 9.8 x 0.5
= 24.5N

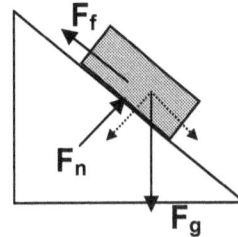

The force of friction up the plane = μ $\mathbf{F_n}$ = 0.5 x 42.6 = 21.3N

Thus the net force on the box acts down the plane and is equal to 24.5 – 21.3 = 3.2N

Skill 7.4   Solve statics problems involving torques

Torque is rotational motion about an axis. It is defined as $\tau = L \times F$, where L is the lever arm. The length of the lever arm is calculated by measuring the perpendicular line drawn from the line of force to the axis of rotation.

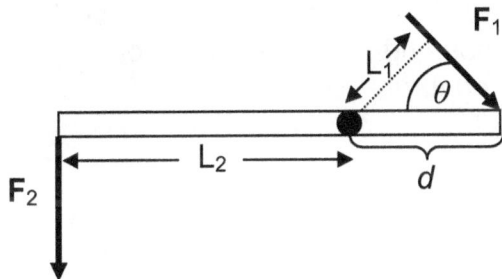

By convention, torques that act in a clockwise direction are considered negative and those in a counterclockwise direction are considered positive. In order for an object to be in equilibrium, the sum of the torques acting on it must be zero.

The equation that would put the above figure in equilibrium is $F_1 L_1 = -F_2 L_2$ (please note that in this case L₁=dsinθ).

Examples:
Some children are playing with the spinner below, when one young boy decides to pull on the spinner arrow in the direction indicated by $\mathbf{F_1}$. How much torque does he apply to the spinner arrow?

$\tau = L \times F$, but L=0 directly through the perpendicular the pivot point is 0). torque to the

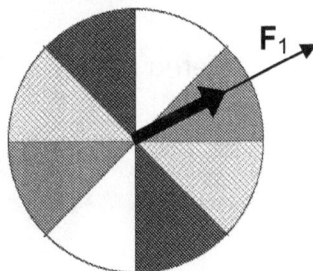

because the line of force goes axis of rotation (i.e. the distance from the line of force to Therefore the boy applies no spinner.

Problem: In the system diagrammed below, find out what the magnitude of F_1 must be in order to keep the system in equilibrium.

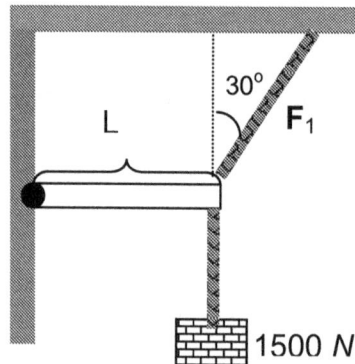

Solution: For an object to be in equilibrium the forces acting on it must be balanced. This applies to linear as well as rotational forces known as moments or torques.

$$F_1 L = -F_2 L$$

$$F_{1x} L + F_{1y} L = -(-1500L) \ldots \text{but } F_{1x} = 0 *$$

$$F_{1y} L = 1500L$$

$$F_1 \cos 30 L = 1500L$$

$$(0.866)F_1 = 1500$$

$$F_1 = 1732N$$

The effect of F_{1x} = 0 because that portion of the force goes right through the pivot and causes no torque.

Things to remember:

1. When writing the equation for a body at equilibrium, the point chosen for the axis of rotation is arbitrary.

2. The center of gravity is the point in an object where its weight can be considered to act for the purpose of calculating torque.

3. The lever arm, or moment arm, of a force is calculated as the **perpendicular** distance from the line of force to the pivot point/axis of rotation.

4. Counterclockwise torques are considered positive while clockwise torques are considered negative.

COMPETENCY 8.0   Understand and apply the laws of motion (including relativity)

Skill 8.1   Describe the characteristics of each of Newton's laws of motion, and analyze their applications

Newton's first law of motion: "An object at rest tends to stay at rest and an object in motion tends to stay in motion with the same speed and in the same direction unless acted upon by an unbalanced force". Prior to Newton's formulation of this law, being at rest was considered the natural state of all objects, because at the earth's surface we have the force of gravity working at all times which causes nearly any object put into motion to eventually come to rest. Newton's brilliant leap was to recognize that an unbalanced force changes the motion of a body, whether that body begins at rest or at some non-zero speed.

We experience the consequences of this law everyday. For instance, the first law is why seat belts are necessary to prevent injuries. When a car stops suddenly, say by hitting a road barrier, the driver continues on forward until acted upon by a force. The seat belt provides that force and distributes the load across the whole body rather than allowing the driver to fly forward and experience the force against the steering wheel.

<u>Example:</u> A skateboarder is riding her skateboard down a road. The skateboard has a constant speed of 5 m/s. Then the skateboard hits a rock and stops suddenly. Since the rider has nothing to stop her when the skateboard stops, she will continue to travel at 5 m/s until she hits the ground.

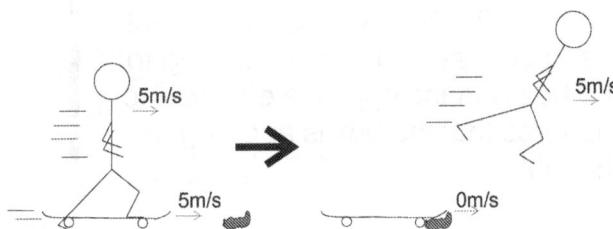

Newton's second law of motion: "The acceleration of an object as produced by a net force is directly proportional to the magnitude of the net force, in the same direction as the net force, and inversely proportional to the mass of the object". In the equation form, it is stated as $F = ma$, force equals mass times acceleration. It is important to remember that this is the net force and that forces are vector quantities. Thus if an object is acted upon by 12 forces that sum to zero, there is no acceleration. Also, this law embodies the idea of inertia as a consequence of mass. For a given force, the resulting acceleration is proportionally smaller for a more massive object because the larger object has more inertia.

Example:
A ball is dropped from a building. The mass of the ball is 2 kg. The acceleration of the object is 9.8 m/s² (gravitational acceleration). Therefore, the force acting on the ball is

$F = ma \Rightarrow F = 2 \text{ kg} \times 9.8 \text{ m/s}^2 \Rightarrow F = 19.6 \text{ N}$

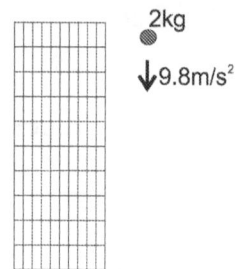

Newton's third law of motion: "For every action, there is an equal and opposite reaction". This statement means that in every interaction, there is a pair of forces acting on the two interacting objects. The size of the force on the first object equals the size of the force on the second object. The direction of the force on the first object is opposite to the direction of the force on the second object.

Example: A box is sitting on a table. The mass of the box is 4 kg. Because of the effects of gravity, the box is applying a force of 39.2 N on the table. The table does not break or shift under the force of the box. This implies that the table is applying a force of 39.2 N on the box. Note that the force that the table is applying to the box is in the opposite direction to the force that the box is applying to the table.

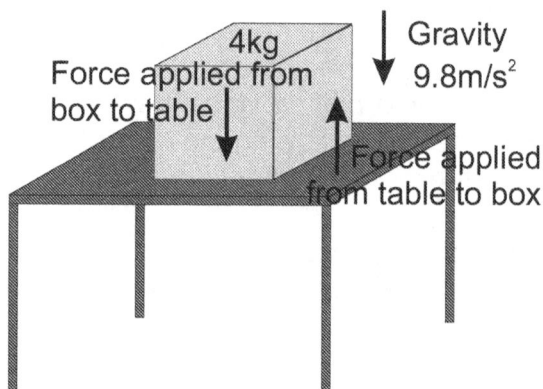

Here are a few more examples:

1. The propulsion/movement of fish through water: A fish uses its fins to push water backwards. The water pushes back on the fish. Because the force on the fish is unbalanced the fish moves forward.

2. The motion of car: A car's wheels push against the road and the road pushes back. Since the force of the road on the car is unbalanced the car moves forward.

3. Walking: When one pushes backwards on the foot with the muscles of the leg, the floor pushes back on the foot. If the forces of the leg on the foot and the floor on the foot are balanced, the foot will not move and the muscles of the body can move the other leg forward.

Skill 8.2 Apply Newton's laws of motion to solve problems

Newton's laws of motion can be used together or separately to analyze a variety of physical situations. Simple examples are provided below and the importance of each law is highlighted.

Problem:
A 10 kg object moves across a frictionless surface at a constant velocity of 5 m/s. How much force is necessary to maintain this speed?

Solution:
Both Newton's first and second laws can help us understand this problem. First, the first law tells us that this object will continue its state of uniform speed in a straight line (since there is no force acting upon it). Additionally, the second law tells that because there is no acceleration (velocity is constant), no force is required. Thus, zero force is necessary to maintain the speed of 5 m/s.

Problem:
A car is driving down a road at a constant speed. The mass of the car is 400 kg. The force acting on the car is 4000 N and the force is in the same direction as the acceleration. What is the acceleration of the car?

Solution: $F = ma \Rightarrow a = \dfrac{F}{m} \Rightarrow a = \dfrac{4000N}{400kg} \Rightarrow a = 10 m/s^2$

Problem:
For the arrangement shown, find the force necessary to overcome the 500 N force pushing to the left and move the truck to the right with an acceleration of 5 m/s^2.

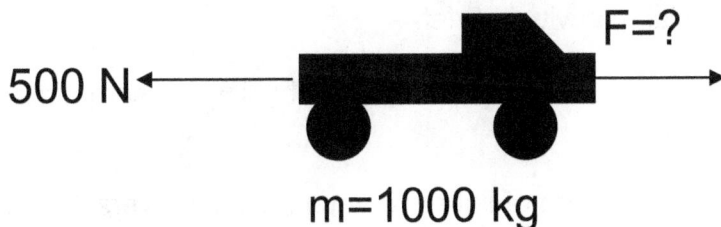

$$500\ N \longleftarrow \qquad F=? \longrightarrow$$

$$m=1000\ kg$$

Solution:
The net force on the truck acting towards the right is F − 500N.
Using Newton's second law, F-500N = 1000kg x 5 m/s^2.
Solving for F, we get F = 5500 N.

Problem:
An astronaut with a mass of 95 kg stands on a space station with a mass of 20,000 kg. If the astronaut is exerting 40 N of force on the space station, what is the acceleration of the space station and the astronaut?

Solution
To find the acceleration of the space station, we can simply apply Newton's second law:

$$A_s = \frac{F}{m_s} = \frac{40N}{20000kg} = 0.002\,{}^m\!/_{s^2}$$

To find the acceleration of the astronaut, we must first apply Newton's third law to determine that the space station exerts an opposite force of -40 N on the astronaut. Here the minus sign simply denotes that the force is directed in the opposite direction. We can then calculate the acceleration, again using Newton's second law:

$$A_a = \frac{F}{m_a} = \frac{-40N}{95kg} = -0.42\,{}^m\!/_{s^2}$$

Skill 8.3   Understand the implications of special relativity for the laws of motion

Postulates of special relativity
Einstein's theory of special relativity built upon the foundation of Galilean relativity and incorporated an analysis of electromagnetics. Galilean relativity is based upon the concept that the laws of physics are invariant with respect to inertial frames of reference. In the context of classical electrodynamics, it is found that Maxwell's equations do not imply any variation in the speed of light with respect to the relative motion of the source and observer. As a result, the second fundamental postulate of special relativity is a statement of the invariance of the speed of light, c, with respect to inertial reference frames.

The postulates of special relativity in brief:

1. **Special principle of relativity:** The laws of physics are same in all inertial frames of reference.

2. **Invariance of c:** The speed of light in a vacuum is a universal constant for all observers, regardless of the inertial frame of reference or the relative motion of the source.

Force and acceleration

One of the implications of these postulates is that the relativistic mass m of a particle is dependent upon its velocity v. This relationship between the rest mass m_0 and the relativistic mass is expressed through the Lorentz factor γ.

$$m = \gamma m_0 = \frac{m_0}{\sqrt{1 - \frac{v^2}{c^2}}}$$

This has a particular effect on the relationship of force F and acceleration a, which is generally expressed using the momentum p.

$$F = \frac{\partial p}{\partial t} = \frac{\partial (mv)}{\partial t}$$

In the context of special relativity, where the mass is dependent on the velocity (which is a function of time), the above equation does not simplify to F = ma, as it does in Newtonian mechanics.

$$F = v\frac{\partial m}{\partial t} + ma = m_0\left(v\frac{\partial \gamma}{\partial t} + \gamma a\right)$$

Since the Lorentz factor increases asymptotically towards infinity as the speed of the particle or object approaches the speed of light, the force required to accelerate the particle also approaches infinity. It would then require infinite energy to accelerate an object to the speed of light, thus making c the "speed limit of the universe." In cases where the speed of the object is significantly less than c, the Lorentz factor is close to unity, thus making Newtonian mechanics an accurate approximation. (This concept is similar to the correspondence principle of quantum mechanics; in this case, relativistic mechanics becomes Newtonian mechanics in the limit as v goes to zero.)

Velocity

As mentioned, the postulates of special relativity seem to imply a universal speed limit. Thus, if this is to be the case in all inertial frames of reference, the observed speed of an object cannot be greater than the speed of light regardless of the velocities of the object and inertial frame of reference as measured in any other inertial frame of reference. As a result, although a particular frame of reference may be moving in one direction at speed w close to light speed and a particular object is moving in the opposite direction at a speed v, likewise near c, the speed of the object as measured from the reference frame *cannot* simply be the sum of v and w. Instead, special relativity uses a different approach to the calculation resulting in a relative velocity given by the following formula.

$$v' = \frac{v + w}{1 + vw/c^2}$$

Conservation of energy

In classical mechanics the behavior of particles or objects can be determined through the application of the conservation of energy and the conservation of momentum. Although, in this case, these two are seemingly independent concepts, in the case of relativistic mechanics they are shown to be interdependent. This interdependence results from the relationship of mass and energy that is derived from the application of the postulates of special relativity.

$$E^2 = \left(mc^2\right)^2 + \left(pc\right)^2$$

Thus, mass and energy (and, therefore, energy and momentum) are interdependent characteristics of the particle or object. The conservation laws for classical momentum and energy are then joined into a more general expression of the conservation of energy.

COMPETENCY 9.0   Understand uniform circular motion and simple harmonic motion, and solve problems involving these types of motion

Skill 9.1   Apply vector analysis to describe uniform circular motion

Speed remains constant in uniform circular motion. However, since the object is following a curve, the direction of the velocity is changing. Therefore, the object is accelerating. This acceleration is called **centripetal acceleration**.

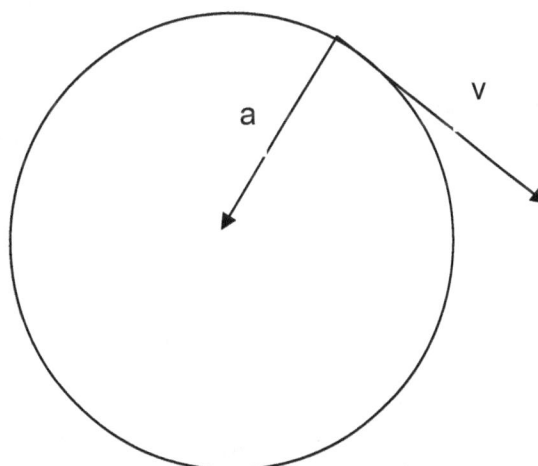

As seen in the above figure, the velocity of the object traveling in a circle is a tangent to the circle. The acceleration of the object is directed towards the center of the circle. Therefore, the object follows a curve.

In order to analyze a circle, it is helpful to use polar coordinates as reference.

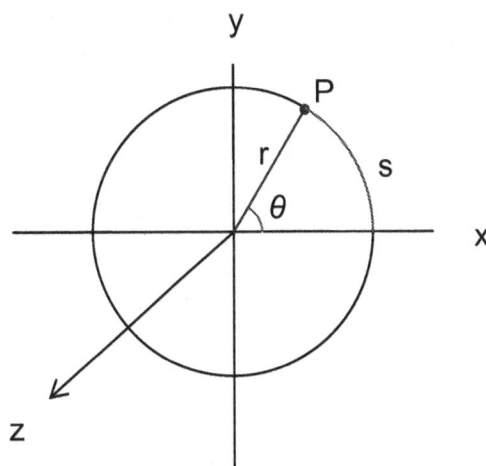

Using polar coordinates (r, θ), Point P on the circle can be defined by the distance from the origin r and the angle with respect to the x-axis θ.

The arc length s can be found using the following equation:

$$s = r\theta$$

where r is the radius and θ is the angular position.

The angular position θ is measure in radians. Note that if the angular position is given in degrees, they must be converted into radians first.

$$1 \text{ revolution} = 360° = 2\pi \text{ radians}$$

Angular velocity

Angular velocity is the angular displacement in a given time period. The instantaneous angular velocity can be found using the equation

$$\omega = \frac{d\theta}{dt}$$

where ω is the angular velocity, θ is the angular position and t is the time.

The relationship between linear velocity (v) and angular velocity (ω)

Imagine there are two revolving bodies that complete one revolution in the same amount of time. However the radius of the orbit of one of the bodies is twice that of the other. It is obvious that the object that has the larger orbit has a greater velocity as it needs to cover a larger distance in the same amount of time. Conversely, the angular velocity of the two bodies is the same as they sweep out equal angles in equal time. Therefore, linear velocity (v) is a function of the radius of a circle (r) for a constant angular velocity (ω).

If $s = r\theta$ then differentiating for time $\dfrac{ds}{dt} = \dfrac{dr}{dt}\theta + \dfrac{d\theta}{dt}r$

This equation can be further simplified as the radius of the circle remains constant, therefore $\dfrac{ds}{dt} = \dfrac{d\theta}{dt}r$

Since, $\omega = \dfrac{d\theta}{dt}$ and $\dfrac{ds}{dt} = v$, we have $v = \omega r$

Skill 9.2 **Determine the magnitude and direction of the forces acting on a particle in uniform circular motion**

As explained in the previous section, in uniform circular motion the acceleration (a) is directed toward the center of the circular path and is always perpendicular to the velocity, as shown below:

This centripetal acceleration is mathematically expressed as:

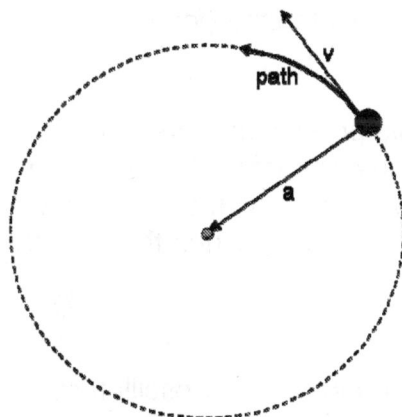

$$a = \frac{v^2}{r} = \frac{4\pi^2 r}{t^2}$$

where r is the radius of the circular path and t is the period of the motion or the time taken for the mass to travel once around the circle. The force (F) experienced by the mass (m) is known as centripetal force and is always directed towards the center of the circular path. It has constant magnitude given by the following equation:

$$F = ma = m\frac{v^2}{r}$$

Problem:
A car of mass 2000 Kg travels round a curve of radius 100m at a speed of 90 Km/h. What is the magnitude of the centripetal force acting on it?

Solution:

$$90Km/h = \frac{90 \times 10^3}{60 \times 60} m/s = 25m/s$$

$$a = \frac{v^2}{r} = \frac{(25)^2}{100} = 6.25m/s^2$$

F = ma = 2000 x 6.25 N = 12500N

Skill 9.3   Understand the relationships among displacement, velocity, and acceleration in simple harmonic motion

Simple harmonic (sinusoidal) motion involves a cyclical exchange of kinetic energy and potential energy as observed in a simple pendulum or a mass on a spring. The relationships among the various parameters of a system displaying simple harmonic motion depends on the type of system being examined. Once the displacement is known, the velocity and acceleration of the object undergoing harmonic motion can be calculated by calculating the first derivative with respect to time (for velocity) or the second derivative with respect to time (for acceleration).

The examples of the mass on a spring and the simple pendulum represent harmonic motion in one dimension and two dimensions, respectively. Each can be treated in a similar manner, although the details vary slightly. It can be shown, however, that the pendulum acts just like the mass on a spring when the displacement angle is small.

Linear oscillator (mass on a spring)
Given the frequency of oscillation f for a mass on a spring (a linear oscillator), along with the concomitant period T = 1/f, the displacement of the mass undergoing harmonic (sinusoidal) motion can be written as follows.

$$x(t) = x_{max} \cos(2\pi f t + \phi) = x_{max} \cos(\omega t + \phi)$$

Alternatively, $2\pi f$ can be written as the angular frequency ω. The coefficient x_{max} is the maximum displacement of the mass, and the term Φ is a phase constant that determines the position of the mass at time t = 0. If x(t) is differentiated once with respect to time, the velocity of the mass is revealed.

$$v(t) = \frac{\partial x(t)}{\partial t} = -\omega x_{max} \sin(\omega t + \phi)$$

Comparing the expressions for displacement and velocity, we can see that the velocity is maximized when the displacement is zero (all kinetic energy), and the displacement is maximized when the velocity is zero (all potential energy); that is, the displacement and velocity are 90^0 out of phase. The acceleration can be calculated by differentiating v(t) with respect to time.

$$a(t) = \frac{\partial v(t)}{\partial t} = -\omega^2 x_{max} \cos(\omega t + \phi) = -\omega^2 x(t)$$

The acceleration, as shown above, is in phase with the displacement. Applying Newton's second law of motion leads to Hooke's law, which relates the restoring force on the mass to the displacement x(t).

$$F = ma = -m\omega^2 x(t)$$

This equation may be expressed in terms of the so-called spring constant k, which is defined as $m\omega^2$.

$$F = ma = -k\,x(t)$$

Simple pendulum

The simple pendulum model provides a reasonably accurate representation of pendulum motion, especially in the case of small angular amplitude.

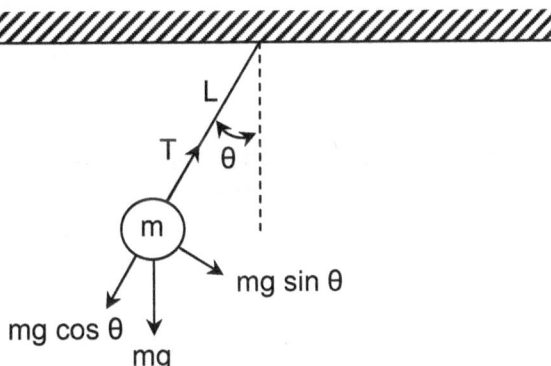

The restoring force F in pendulum motion is expressed as a component of the gravitational force mg perpendicular to the length of the string and is given by

$$F = -mg\sin\theta$$

The negative sign results from the force having a direction opposite to the displacement. The tension T on the string and the portion of the gravitational force in the opposite direction of T cancel one another.

In the case of small θ, $\sin\theta$ is approximately equal to θ. The arc length traveled by the pendulum, s, is equal to the product of the length L of the string and the angle θ. Thus, the following expression can be derived.

$$F \approx -mg\theta = -mg\frac{s}{L} = -\frac{mg}{L}s$$

In the above equation, the force is shown to be of the same form as the linear harmonic oscillator, having, in this case, a "spring constant" of mg/L. As a result, the expressions found for the displacement, velocity and acceleration in the case of the linear oscillator can also be used here (in the case of small θ), where the frequency ω is replaced as follows.

$$\omega = \sqrt{\frac{g}{L}}$$

Skill 9.4 **Apply the principles of simple harmonic motion to solve problems involving oscillatory phenomena**

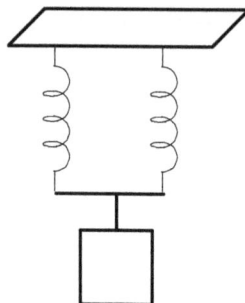

The above diagram is an example of a **Hookean system** (a spring, wire, rod, etc.) where the spring returns to its original configuration after being displaced and then released. When the spring is stretched a distance x, the restoring force exerted by the spring is expressed by Hooke's Law

$$F = -kx$$

The minus sign indicates that the restoring force is always opposite in direction to the displacement. The variable k is the spring constant and measures the stiffness of the spring in N/m.

The period of simple harmonic motion for a Hookean spring system is dependent upon the mass (m) of the spring and the stiffness of the spring (k) and is given by

$$T = \frac{1}{f} = \frac{1}{\frac{\omega}{2\pi}} = \frac{2\pi}{\omega} = \frac{2\pi}{\sqrt{\frac{k}{m}}} = 2\pi\sqrt{\frac{m}{k}}$$

Problem:
Each spring in the above diagram has a stiffness of $k = 20N/m$. The mass of the object connected to the spring is 2 kg. Ignoring friction forces, find the period of motion.

Solution:
Utilizing Hooke's Law, the net restoring force on the spring would be

$$F = -(20N/m)x - (20N/m)x = -(40N/m)x$$

Comparison with $F = -kx$ shows the equivalent k to be 40 N/m.
Using the above formula for our problem, we have

$$T = 2\pi\sqrt{\frac{m}{k}} = 2\pi\sqrt{\frac{2kg}{40N/m}} = 2(3.14)\sqrt{0.05} = 1.4s$$

COMPETENCY 10.0   Understand the concepts of energy, work, and power, and the principles of conservation of energy and momentum

Skill 10.1   Analyze mechanical systems in terms of work, power, and energy

In physics, work is defined as force times distance $W = F \cdot s$. Work is a scalar quantity, it does not have direction, and it is usually measured in Joules ($N \cdot m$). It is important to remember, when doing calculations about work, that the only part of the force that contributes to the work is the part that acts in the direction of the displacement. Therefore, sometimes the equation is written as $W = F \cdot s \cos\theta$, where θ is the angle between the force and the displacement.

Problem:
A man uses 6N of force to pull a 10kg block, as shown below, over a distance of 3 m. How much work did he do?

Solution:

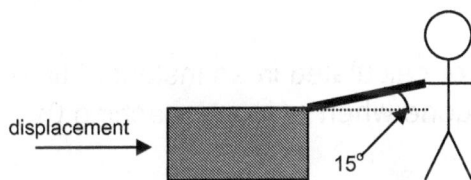

$$W = F \cdot s \cos\theta$$
$$W = 6 \cdot 3 \cos 15 = 17.4 J$$

Notice that you did not actually need the mass of the box in order to calculate the work done.

The power expended by a system can be defined as either the rate at which work is done by the system or the rate at which energy is transferred from it. There are many different measurements for power, but the one most commonly seen in physics problems is the Watt which is measured in Joules per second. Another commonly discussed unit of power is horsepower, and 1hp=746 W.

The **average power** of a system is defined as the work done by the system divided by the total change in time:

$\overline{P} = \dfrac{W}{\Delta t} \Rightarrow$ Where $\overline{P}$ = average power, W = work and Δt = change in time

The average power can also be written in terms of energy transfer $\Rightarrow \overline{P} = \dfrac{\Delta E}{\Delta t}$ and used the same way that the equation for work is used.

Problem:
A woman standing in her 4th story apartment raises a 10kg box of groceries from the ground using a rope. She is pulling at a constant rate, and it takes her 5 seconds to raise the box one meter. How much power is she using to raise the box?

Solution:

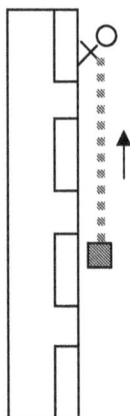

$$P = W/t$$

$$P = \frac{F \cdot s}{t} = \frac{mgh}{t} = \frac{10*9.8*1}{5} = 19.6W$$

Notice that because she is pulling at a constant rate, you don't need to know the actual distance she has raised the box. 2 meters in 10 seconds would give you the same result as 5 meters in 25 seconds.

Instantaneous power is the power measured or calculated in an instant of time. Since instantaneous power is the rate of work done when Δt is approaching 0s, the power is then written in derivate form:

$$P = \frac{dW}{dt} \Rightarrow \text{Where } P = \text{average power, } dW = \text{work and } dt = \text{change in time.}$$

Since $W = Fs\cos\phi$, for a constant force the above equation can be written as:

$$P = \frac{dW}{dt} \Rightarrow P = \frac{d(Fs\cos\phi)}{dt} \Rightarrow P = \frac{(F\cos\phi)ds}{dt} \Rightarrow P = F\cos\phi\left(\frac{ds}{dt}\right) \Rightarrow P = Fv\cos\phi$$

where v is the velocity of the object.

Skill 10.2   Use the concept of conservation of energy to solve problems

According to the concept of conservation of energy, the energy in an isolated system remains the same although it may change in form. For instance, potential energy can become kinetic energy and kinetic energy, depending on the system, can become thermal or heat energy. Solving energy conservation problems depends on knowing the types of energy one is dealing with in a particular situation and assuming that the sum of all the different types of energy remains constant. Below we will discuss several different examples.

Example:

A rollercoaster at the top of a hill has a certain potential energy that will allow it to travel down the track at a speed based on its potential energy and friction with the track itself. At the bottom of the hill, when it has reached a stop, its potential energy is zero and all of the energy has been transferred from potential energy to kinetic energy (movement) and thermal energy (heat derived from friction). The equation below describes the relationship between potential energy and other forms of energy in this case:

Potential Energy = Kinetic energy(movement) + heat energy(friction)

Problem:

A skier travels down a ski slope with negligible friction. He begins at 100 meters in height, drops to a much lower level and ends at 90 meters in height. What is the skier's velocity at the 90 meter height?

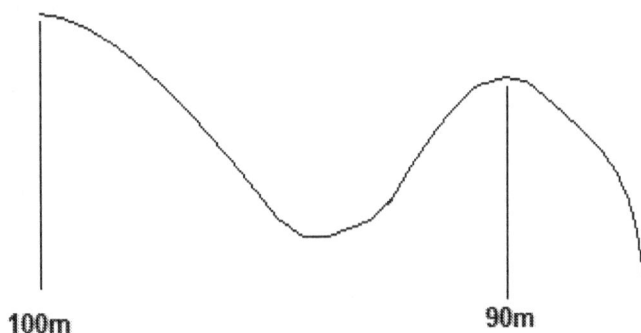

100m   90m

The skier's initial energy is only potential energy and is given by mgh = 100mg where m is the mass of the skier. The skier's final energy is the sum of his potential and kinetic energies and is given by $mgh + 1/2mv^2 = 90mg + 1/2mv^2$ where v is the skier's velocity. Using the principle of conservation of energy we know that the initial and final energies must be equal. Hence: $100mg = 90mg + 1/2mv^2$

Thus, $1/2mv^2 = 10mg$; $1/2v^2 = 10g$; $v = 14m/s$ since ($g=9.81m/s^2$)

Problem:

d. A pebble weighing 10 grams is placed in a massless frictionless sling shot, with a spring constant of 200N/m, that is stretched back 0.5 meters. What is the total energy of the system before the pebble is released? What is the final height of the pebble if it is shot straight up and the effects of air resistance are negligible?

Solution:

If the initial height of the pebble is h = 0, total energy is given by
$E = ½ mv^2 + mgh + ½ kx^2 = 0 + 0 + 0.5(200)(0.5)^2 = 25$ Joules

At its final height, the velocity of the pebble will be zero. Since
$E = ½ mv^2 + mgh + ½ kx^2$, from the principle of conservation of energy
25 Joules = 0 + 0.010kg (9.81)h + 0 and h = 255 m

Skill 10.3   Determine power, mechanical advantage, and efficiency as they relate to work and energy in simple machines

Work is defined as the product of force and distance and **power** as the rate at which work is done. A simple machine, consisting of a single mechanical device, is often used to "make work easier". A machine can be considered in terms of input force and output force. Additionally, when considering the distances over which those forces are applied there is input and output work.

In the ideal case of a frictionless work, the input work equals the output work. However, new way machines can effectively make tasks easier is by requiring less force. For example, some work is done by a force F_1 over a distance D_1. The same work can be accomplished by a lesser force $F_2 < F_1$ over a greater distance $D_2 > D_1$.

$$\text{Input Work} = \text{Output Work}$$
$$(\text{Input Force})(\text{Distance}) = (\text{Output Force})(\text{Distance})$$
$$F_1 D_1 = F_2 D_2$$

Reduced force over a longer distance is the concept behind an inclined plane. Other simple machines operate by different principles that "make work easier" such as transferring a force from one place to another, changing the direction of a force, increasing the magnitude of a force, or increasing the distance or speed of a force. The six types of simple machines:

- Inclined Plane
- Lever
- Wheel and Axle
- Pulley
- Wedge
- Screw

The efficiency of a machine is percentage of input work that converted to usable output work or the ratio of work output to work input.

$$(2)\ \text{Efficiency} = \frac{\text{Work Output}}{\text{Work Input}}$$

The efficiency of a machine can never equal 100% because some output work is lost to friction.

Skill 10.4   Use the concept of conservation of momentum to solve problems

The **impulse-momentum theorem** states that any impulse acting on a system changes the momentum of that system. When considering the impulse-momentum theorem, there are several factors that need to be taken into account. The first factor is that momentum is a vector quantity $p = m \cdot v$. It has both magnitude and direction. Therefore, any action that causes either the speed or the direction of an object to change causes a change in its momentum. An impulse is defined as a force acting over a period of time (integral of force over time), and any impulse acting on the system is equivalent to a change in its momentum, as you can see from the equations below:

$$F = m \cdot a \to F = m \cdot \frac{\Delta v}{t} \to F \cdot t = m \cdot \Delta v$$

i.e. Forces acting over time cause a change in momentum.

Sample Problems:

1. A 1 kg ball is rolled towards a wall at 4 m/s. It hits the wall, and bounces back off the wall at 3 m/s.

    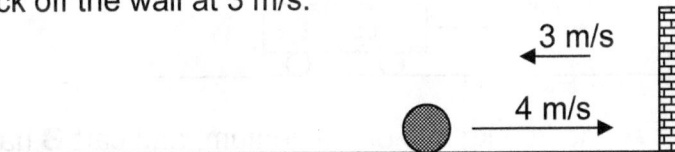

    a. What is the change in velocity?

        The velocity goes from +4m/s to –3m/s, a net change of -7m/s.

    b. At what point does the impulse occur?

        The impulse occurs when the ball hits the wall.

2. A 30kg woman is in a car accident. She was driving at 50m/s when she had to hit the brakes to avoid hitting the car in front of her.

    a. The automatic tensioning device in her seatbelt slows her down to a stop over a period of one half second. How much force does it apply?
        $$F = m \cdot \frac{\Delta v}{t} \to F = 30 \cdot \frac{50}{.5} = 3000 N$$

    b. If she hadn't been wearing a seatbelt, the windshield would have stopped her in .001 seconds. How much force would have been applied there?
        $$F = m \cdot \frac{\Delta v}{t} \to F = 30 \cdot \frac{50}{.001} = 1500000 N$$

The **law of conservation of momentum** states that the total momentum of an *isolated system* (not affected by external forces and not having internal dissipative forces) always remains the same. For instance, in any collision between two objects in an isolated system, the total momentum of the two objects after the collision will be the same as the total momentum of the two objects before the collision. In other words, any momentum lost by one of the objects is gained by the other.

A collision may be **elastic** or **inelastic**. In a totally elastic collision, the kinetic energy is conserved along with the momentum. In a totally inelastic collision, on the other hand, the kinetic energy associated with the center of mass remains unchanged but the kinetic energy relative to the center of mass is lost. An example of a totally inelastic collision is one in which the bodies stick to each other and move together after the collision. Most collisions are neither perfectly elastic nor perfectly inelastic and only a portion of the kinetic energy relative to the center of mass is lost.

Example: Inelastic Collision

Imagine two carts rolling towards each other as in the diagram below

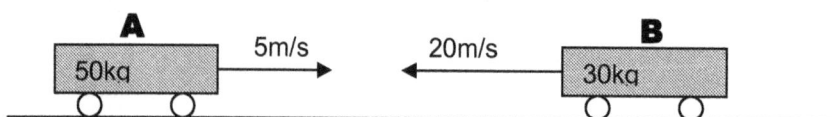

Before the collision, cart **A** has 250 kg m/s of momentum, and cart **B** has –600 kg m/s of momentum. In other words, the system has a total momentum of –350 kg m/s of momentum.

After the collision, the two carts stick to each other, and continue moving. How do we determine how fast, and in what direction, they go?

We know that the new mass of the cart is 80kg, and that the total momentum of the system is –350 kg m/s. Therefore, the velocity of the two carts stuck together must be $\dfrac{-350}{80} = -4.375\,m/s$

Conservation of momentum works the same way in two dimensions, the only change is that you need to use vector math to determine the total momentum and any changes, instead of simple addition.

Example: Elastic Collision

Imagine a pool table like the one below. Both balls are 0.5 kg in mass.

Before the collision, the white ball is moving with the velocity indicated by the solid line and the black ball is at rest.

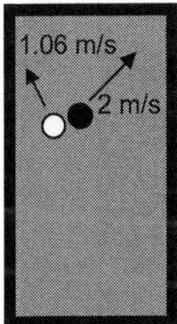

After the collision the black ball is moving with the velocity indicated by the dashed line (a 135° angle from the direction of the white ball).

With what speed, and in what direction, is the white ball moving after the collision?

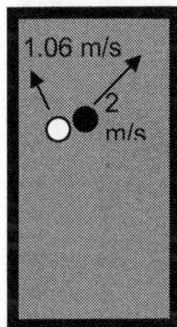

$$p_{white\,/\,before} = .5 \cdot (0,3) = (0,1.5) \quad p_{black\,/\,before} = 0 \quad p_{total\,/\,before} = (0,1.5)$$

$$p_{black\,/\,after} = .5 \cdot (2\cos 45, 2\sin 45) = (0.71, 0.71)$$

$$p_{white\,/\,after} = (-0.71, 0.79)$$

i.e. the white ball has a velocity of

$v = \sqrt{(-.71)^2 + (0.79)^2} = 1.06 m/s$, and is moving at an angle

of $\theta = \tan^{-1}\left(\dfrac{0.79}{-0.71}\right) = -48°$ from the horizontal

COMPETENCY 11.0 **Understand the dynamics of rotational motion and the properties of fluids**

Skill 11.1 **Apply the law of conservation of angular momentum**

One important difference in the equations relates to the use of mass in rotational systems. In rotational problems, not only is the mass of an object important but also its location. In order to include the spatial distribution of the mass of the object, a term called **moment of inertia** is used, $I = m_1 r_1^2 + m_2 r_2^2 + \cdots + m_n r_n^2$. The moment of inertia is always defined with respect to a particular axis of rotation.

Example:

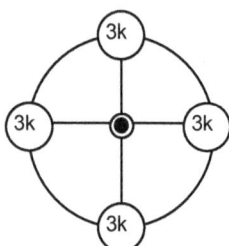

If the radius of the wheel on the left is 0.75m, what is its moment of inertia about an axis running through its center perpendicular to the plane of the wheel?

$$I = 3 \cdot 0.75^2 + 3 \cdot 0.75^2 + 3 \cdot 0.75^2 + 3 \cdot 0.75^2 = 6.75$$

Note: $I_{Sphere} = \frac{2}{5} mr^2$, $I_{Hoop/Ring} = mr^2$, $I_{disk} = \frac{1}{2} mr^2$

The rotational analog of Newton's second law of motion is given in terms of **torque** τ, moment of inertia I, and angular acceleration α: $\tau = I\alpha$

where the torque τ is the rotational force on the body. In simple terms, the torque τ produced by a force F acting at a distance r from the point of rotation is given by the product of r and the component of the force that is perpendicular to the line joining the point of rotation to the point of action of the force.

A concept related to the moment of inertia is the **radius of gyration** (*k*), which is the average distance of the mass of an object from its axis of rotation, i.e., the distance from the axis where a point mass *m* would have the same moment of inertia.

$$k_{Sphere} = \sqrt{\frac{2}{5}} r, \, k_{Hoop/Ring} = r, \, k_{disk} = \frac{r}{\sqrt{2}}.$$

As you can see

$$I = mk^2$$

This is analogous to the concept of center of mass, the point where an equivalent mass of infinitely small size would be located, in the case of linear motion.

Angular momentum (L), and **rotational kinetic energy (KE$_r$),** are therefore defined as follows: $L = I\omega, \quad KE_r = \frac{1}{2}I\omega^2$

Skill 11.2 **Apply the concepts of center of mass, moment of inertia, and rotational kinetic energy to analyze the rotational motion of an object**

Unless a net torque acts on a system, the angular momentum remains constant in both magnitude and direction. This can be used to solve many different types of problems including ones involving satellite motion.

Example:
A planet of mass m is circling a star in an orbit like the one below. If its velocity at point A is 60,000m/s, and $r_B = 8\ r_A$, what is its velocity at point B?

Solution:

$$I_B\omega_B = I_A\omega_A$$

$$mr_B^2\omega_B = mr_A^2\omega_A$$

$$r_B^2\omega_B = r_A^2\omega_A$$

$$r_B^2\frac{v_B}{r_B} = r_A^2\frac{v_A}{r_A}$$

$$r_Bv_B = r_Av_A$$

$$8r_Av_B = r_Av_A$$

$$v_B = \frac{v_A}{8} = 7500m/s$$

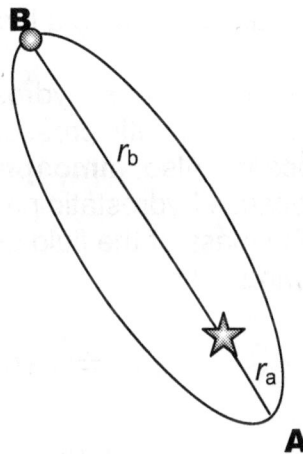

Example:

A uniform ball of radius r and mass m starts from rest and rolls down a frictionless incline of height h. When the ball reaches the ground, how fast is it going?

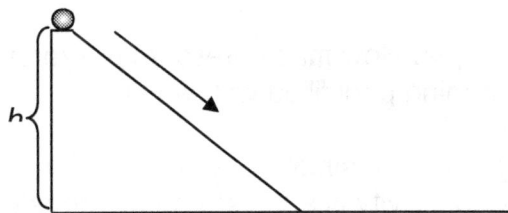

Solution:

$$PE_{initial} + KE_{rotational\,/\,initial} + KE_{linear\,/\,initial} = PE_{final} + KE_{rotational\,/\,final} + KE_{linear\,/\,final}$$

$$mgh + 0 + 0 = 0 + \frac{1}{2}I\omega_{final}^{2} + \frac{1}{2}mv_{final}^{2} \rightarrow mgh = \frac{1}{2}\cdot\frac{2}{5}mr^{2}\omega_{final}^{2} + \frac{1}{2}mv_{final}^{2}$$

$$mgh = \frac{1}{5}mr^{2}(\frac{v_{final}}{r})^{2} + \frac{1}{2}mv_{final}^{2} \rightarrow mgh = \frac{1}{5}mv_{final}^{2} + \frac{1}{2}mv_{final}^{2}$$

$$gh = \frac{7}{10}v_{final}^{2} \rightarrow v_{final} = \sqrt{\frac{10}{7}gh}$$

Skill 11.3   Apply the concepts of force, pressure, and density

Pressure is the force applied per unit area on a surface and is measured in the units N/m^{2} in the SI system. Density is the mass per unit volume and is measured in the SI units kg/m^{3}. Pressure and density (in place of force and mass) are useful concepts when we are dealing with an extended mass of substance such as a fluid as opposed to a lump of matter such as a solid ball.

The weight of a column of fluid creates hydrostatic pressure. Common situations in which we might analyze hydrostatic pressure include tanks of fluid, a swimming pool, or the ocean. Also, **atmospheric pressure** is an example of hydrostatic pressure. Because hydrostatic pressure results from the force of gravity interacting with the mass of the fluid or gas, for an incompressible fluid it is governed by the following equation:

$$P = \rho gh$$

where P=hydrostatic pressure
ρ=density of the fluid
g=acceleration of gravity
h=height of the fluid column

Example: How much pressure is exerted by the water at the bottom of a 5 meter swimming pool filled with water?

Solution: We simply use the equation from above, recalling that the acceleration due to gravity is 9.8m/s² and the density of water is 1000 kg/m³.

$$P = \rho gh = 1000\frac{kg}{m^{3}} \times 9.8\frac{m}{s^{2}} \times 5m = 49,000\,Pa = 49\,kPa$$

Skill 11.4   Use Archimedes' and Bernoulli's principles to analyze properties of fluids

Archimedes' Principle
Archimedes discovered through observation that, for an object in a fluid, "the upthrust is equal to the weight of the displaced fluid" and the weight of displaced fluid is directly proportional to the volume of displaced fluid. The second part of his discovery is useful when we want to, for instance, determine the volume of an oddly shaped object. We determine its volume by immersing it in a graduated cylinder and measuring how much fluid is displaced. We explore his first observation in more depth below.

Today, we call Archimedes' "upthrust" **buoyancy**. Buoyancy is the force produced by a fluid on a fully or partially immersed object. The buoyant force ($F_{buoyant}$) is found using the following equation:

$$F_{bouyant} = \rho V g$$

where ρ = density of the fluid
V = volume of fluid displaced by the submerged object
g = the acceleration of gravity

Notice that the buoyant force opposes the force of gravity. For an immersed object, a gravitational force (equal to the object's mass times the acceleration of gravity) pulls it downward, while a buoyant force (equal to the weight of the displaced fluid) pushes it upward.

Also note that, from this principle, we can predict *whether* an object will sink or float in a given liquid. We can simply compare the density of the material from which the object is made to that of the liquid. If the material has a lower density, it will float; if it has a higher density it will sink. Finally, if an object has a density equal to that of the liquid, it will neither sink nor float.

Example: Will gold (ρ=19.3 g/cm^3) float in water?

Solution: We must compare the density of gold with that of water, which is 1 g/cm^3.

$$\rho_{gold} > \rho_{water}$$

So, gold will sink in water.

Example: Imagine a 1 m^3 cube of oak (530 kg/m^3) floating in water. What is the buoyant force on the cube and how far up the sides of the cube will the water be?

Solution: Since the cube is floating, it has displaced enough water so that the buoyant force is equal to the force of gravity. Thus the buoyant force on the cube is equal to its weight 1X530X9.8 N = 5194 N.

To determine where the cube sits in the water, we simply the find the ratio of the wood's density to that of the water:

$$\frac{\rho_{oak}}{\rho_{water}} = \frac{530\,^{kg}/_{m^3}}{1000\,^{kg}/_{m^3}} = 0.53$$

Thus, 53% of the cube will be submerged. Since the edges of the cube must be 1m each, the top 0.47m of the cube will appear above the water.

Bernoulli's Principle
The study of moving fluids is contained within fluid mechanics which is itself a component of continuum mechanics. Some of the most important applications of fluid mechanics involve liquids and gases moving in tubes and pipes.

Fluid flow may be laminar or turbulent. One cannot predict the exact path a fluid particle will follow in turbulent or erratic flow. Laminar flow, however, is smooth and each fluid particle follows a continuous path. Lines, know as **streamlines,** are drawn to show the path of a laminar fluid. Streamlines never cross one another and higher fluid velocity is depicted by drawing streamlines closer together.

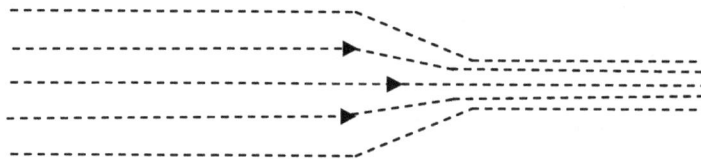

To understand the movement of laminar fluids, one of the first quantities we must define is volumetric flow rate which may have units of gallons per min (gpm), liters/s, cubic feet per min (cfm), gpf, or m^3/s:

$$Q = Av\cos\theta$$

Where Q=volumetric flow rate
A=cross sectional area of the pipe
v=fluid velocity
θ=the angle between the direction of the fluid flow and a vector normal to A

Note that in situations in which the fluid velocity is perpendicular to the cross sectional area, this equation is simply:

$$Q = Av$$

It is also convenient to sometimes discuss mass flow rate ($\dot{m}$), which we can easily find using the density (ρ) of the fluid:

$$\dot{m} = \rho vA = \rho Q$$

Usually, we make an assumption that the fluid is incompressible, that is, the density is constant. Like many commonly used simplifications, this assumption is largely and typically correct though real fluids are, of course, compressible to varying extents. When we do assume that density is constant, we can use conservation of mass to determine that when a pipe is expanded or restricted, the mass flow rate will remain the same.

Let's see how this pertains to an example:

Given conservation of mass, it must be true that:

$$v_1 A_1 = v_2 A_2$$

This is known as the **equation of continuity**. Note that this means the fluid will flow faster in the narrower portions of the pipe and more slowly in the wider regions. An everyday example of this principle is seen when one holds their thumb over the nozzle of a garden hose; the cross sectional area is reduced and so the water flows more quickly.

Much of what we know about fluid flow today was originally discovered by Daniel Bernoulli. His most famous discovery is known as Bernoulli's Principle which states that, if no work is performed on a fluid or gas, an increase in velocity will be accompanied by a decrease in pressure. The mathematical statement of the Bernoulli's Principle for incompressible flow is:

$$\frac{v^2}{2} + gh + \frac{p}{\rho} = \text{constant}$$

where v= fluid velocity
g=acceleration due to gravity
h=height
p=pressure
ρ=fluid density

Though some physicists argue that it leads to the compromising of certain assumptions (i.e., incompressibility, no flow motivation, and a closed fluid loop), most agree it is correct to explain "lift" using Bernoulli's principle. This is because Bernoulli's principle can also be thought of as predicting that the pressure in moving fluid is less than the pressure in fluid at rest.

Thus, there are many examples of physical phenomenon that can be explained by Bernoulli's Principle:

- The lift on airplane wings occurs because the top surface is curved while the bottom surface is straight. Air must therefore move at a higher velocity on the top of the wing and the resulting lower pressure on top accounts for lift.
- The tendency of windows to explode rather than implode in hurricanes is caused by the pressure drop that results from the high speed winds blowing across the outer surface of the window. The higher pressure on the inside of the window then pushes the glass outward, causing an explosion.
- The ballooning and fluttering of a tarp on the top of a semi-truck moving down the highway is caused by the flow of air across the top of the truck. The decrease in pressure causes the tarp to "puff up."
- A perfume atomizer pushes a stream of air across a pool of liquid. The drop in pressure caused by the moving air lifts a bit of the perfume and allows it to be dispensed.

COMPETENCY 12.0   Understand the concept of heat energy and the laws of thermodynamics

Skill 12.1   Solve calorimetry problems involving heat capacity, specific heat, heat of fusion, and heat of vaporization

Heat capacity and phase change

A substance's molar heat capacity is the heat required to change the temperature of one mole of the substance by one degree. Heat capacity has units of joules per mol- kelvin or joules per mol- °C. The two units are interchangeable because we are only concerned with differences between one temperature and another. A Kelvin degree and a Celsius degree are the same size.

The specific heat of a substance (also called specific heat capacity) is the heat required to change the temperature of one gram or kilogram by one degree. Specific heat has units of joules per gram-°C or joules per kilogram-°C.

The temperature of a material rises when heat is transferred to it and falls when heat is removed from it. When the material is undergoing a phase change (e.g. from solid to liquid), however, it absorbs or releases heat without a corresponding change in temperature. The heat that is absorbed or released during phase change is known as latent heat.

A temperature vs. heat graph can demonstrate these relationships visually. One can also calculate the specific heat or latent heat of phase change for the material by studying the details of the graph.

Example: The plot below shows heat applied to 1g of ice at -40C. The horizontal parts of the graph show the phase changes where the material absorbs heat but stays at the same temperature. The graph shows that ice melts into water at 0C and the water undergoes a further phase change into steam at 100C.

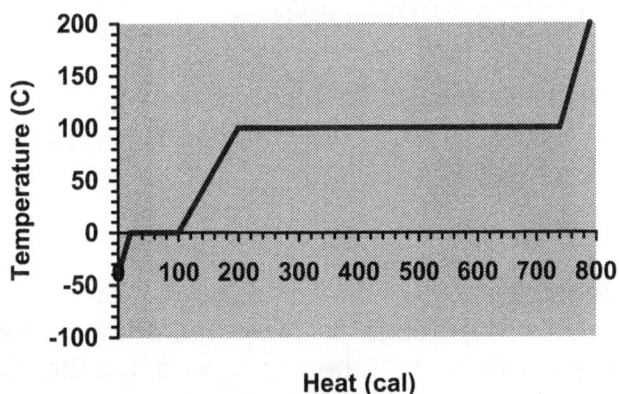

Heat (cal)

The specific heat of ice, water and steam and the latent heat of fusion and vaporization may be calculated from each of the five segments of the graph.

For instance, we see from the flat segment at temperature 0C that the ice absorbs 80 cal of heat. The latent heat L of a material is defined by the equation $\Delta Q = mL$ where ΔQ is the quantity of heat transferred and m is the mass of the material. Since the mass of the material in this example is 1g, the latent heat of fusion of ice is given by $L = \Delta Q / m$ = 80 cal/g.

The next segment shows a rise in the temperature of water and may be used to calculate the specific heat C of water defined by $\Delta Q = mC\Delta T$, where ΔQ is the quantity of heat absorbed, m is the mass of the material and ΔT is the change in temperature. According to the graph, ΔQ = 200-100 =100 cal and ΔT = 100-0=100C. Thus, C = 100/100 = 1 cal/gC.

Problem: The plot below shows the change in temperature when heat is transferred to 0.5g of a material. Find the initial specific heat of the material and the latent heat of phase change.

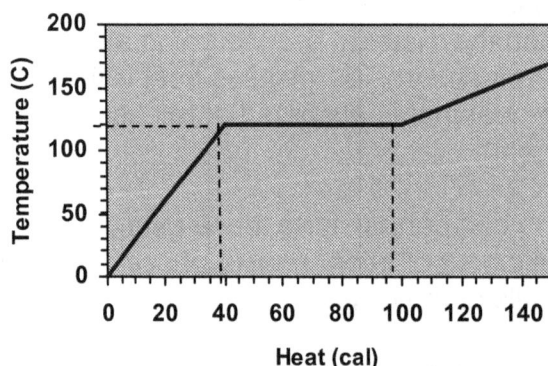

Solution:
Looking at the first segment of the graph, we see that ΔQ = 40 cal and ΔT = 120 C. Since the mass m = 0.5g, the specific heat of the material is given by
$C = \Delta Q / (m\Delta T)$ = 40/(0.5 X120) = 0.67 cal/gC.

The flat segment of the graph represents the phase change. Here
ΔQ = 100 - 40=60 cal. Thus, the latent heat of phase change is given by
$L = \Delta Q / m$ = 60/(0.5) = 120 cal/g.

Heat transfer
The amount of heat transferred by conduction through a material depends on several factors. It is directly proportional to the temperature difference ΔT between the surface from which the heat is flowing and the surface to which it is transferred. Heat flow H increases with the area A through which the flow occurs and also with the time duration t. The thickness of the material reduces the flow of heat.

The relationship between all these variables is expressed as

$$H = \frac{k.t.A.\Delta T}{d}$$

where the proportionality constant k is known as the **thermal conductivity**, a property of the material. Thermal conductivity of a good conductor is close to 1 (0.97 cal/cm.s.^0C for silver) while good insulators have thermal conductivity that is nearly zero (0.0005 cal/cm.s.^0C for wood).

Problem: A glass window pane is 50 cm long and 30 cm wide. The glass is 1 cm thick. If the temperature indoors is 15^0C higher than it is outside, how much heat will be lost through the window in 30 minutes? The thermal conductivity of glass is 0.0025 cal/cm.s.^0C.

Solution: The window has area A = 1500 sq. cm and thickness d = 1 cm. Duration of heat flow is 1800 s and the temperature difference $\Delta T = 15^0C$. Therefore heat loss through the window is given by

$$H = (0.0025 \times 1800 \times 1500 \times 15)/1 = 101250 \text{ calories}$$

The amount of energy radiated by a body at temperature T and having a surface area A is given by the Stefan-Boltzmann law expressed as:

$$I = e\sigma A T^4$$

where I is the radiated power in watts, e (a number between 0 and 1) is the **emissivity** of the body and σ is a universal constant known as **Stefan's constant** that has a value of $5.6703 \times 10^{-8} W/m^2.K^4$. Black objects absorb and radiate energy very well and have emissivity close to 1. Shiny objects that reflect energy are not good absorbers or radiators and have emissivity close to zero.

A body not only radiates thermal energy but also absorbs energy from its surroundings. The net power radiation from a body at temperature T in an environment at temperature T_0 is given by

$$I = e\sigma A(T^4 - T_0^4)$$

Problem: Calculate the net power radiated by a body of surface area 2 sq. m, temperature 30^0C and emissivity 0.5 placed in a room at a temperature of 15^0C.

Solution: $I = 0.5 \times 5.67 \times 10^{-8} \times 2(303^4 - 288^4) = 88 \text{ W}$

Skill 12.2   Analyze methods of heat transfer in practical situations

All heat transfer is the movement of thermal energy from hot to cold matter. This movement down a thermal gradient is a consequence of the second law of thermodynamics. The three methods of heat transfer are listed and explained below.

Conduction: Electron diffusion or photo vibration is responsible for this mode of heat transfer. The bodies of matter themselves do not move; the heat is transferred because adjacent atoms that vibrate against each other or as electrons flow between atoms. This type of heat transfer is most common when two solids come in direct contact with each other. This is because molecules in a solid are in close contact with one another and so the electrons can flow freely. It stands to reason, then, that metals are good conductors of thermal energy. This is because their metallic bonds allow the freest movement of electrons. Similarly, conduction is better in denser solids. Examples of conduction can be seen in the use of copper to quickly convey heat in cooking pots, the flow of heat from a hot water bottle to a person's body, or the cooling of a warm drink with ice.

Convection: Convection involves some conduction but is distinct in that it involves the movement of warm particles to cooler areas. Convection may be either natural or forced, depending on how the current of warm particles develops. Natural convection occurs when molecules near a heat source absorb thermal energy (typically via conduction), become less dense, and rise. Cooler molecules then take their place and a natural current is formed. Forced convection, as the name suggests, occurs when liquids or gases are moved by pumps, fans, or other means to be brought into contact with warmer or cooler masses. Because the free motion of particles with different thermal energy is key to this mode of heat transfer, convection is most common in liquid and gases. Convection can, however, transfer heat between a liquid or gas and a solid. Forced convection is used in "forced air" home heating systems and is common in industrial manufacturing processes. Additionally, natural convection is responsible for ocean currents and many atmospheric events. Finally, natural convection often arises in association with conduction, for instance in the air near a radiator or the water in a pot on the stove.

Radiation: This method of heat transfer occurs via electromagnetic radiation. All matter warmer than absolute zero (that is, all known matter) radiates heat. This radiation occurs regardless of the presence of any medium. Thus, it occurs even in a vacuum. Since light and radiant heat are both part of the EM spectrum, we can easily visualize how heat is transferred via radiation. For instance, just like light, radiant heat is reflected by shiny materials and absorbed by dark materials. Common examples of radiant heat include the way sunlight travels from the sun to warm the earth, the use of radiators in homes, and the warmth of incandescent light bulbs.

PHYSICS   72

Skill 12.3   Apply the first law of thermodynamics in a variety of situations

The internal energy of a material is the sum of the total kinetic energy of its molecules and the potential energy of interactions between those molecules. Total kinetic energy includes the contributions from translational motion and other components of motion such as rotation. The potential energy includes energy stored in the form of resisting intermolecular attractions between molecules.

The enthalpy (H) of a material is the sum of its internal energy and the mechanical work it can do by driving a piston. A change in the enthalpy of a substance is the total energy change caused by adding/removing heat at constant pressure.

The first law of thermodynamics is a general expression of the conservation of energy wherein heat transfer is considered to be a form of energy transfer in addition to work done on or by a system. The internal energy U of a closed system is related to the heat energy Q in the system and the work W done by the system through the following equation.

$$dU = \delta Q - \delta W$$

Thus, the incremental change in the internal energy of the system is equal to the difference between the incremental amount of heat gained by the system and the incremental amount of work done by the system.

The first law of thermodynamics applies in a number of situations. Ideally, in many simple models of various phenomena, undesirable effects such as friction, resistance and absorption are ignored. In reality, however, these are unavoidable consequences of the nature of the phenomena. In the case of mechanical systems. The motion of a moving object can be impeded by the resistance of air, the surfaces of other objects or the viscosity of a liquid. Since energy is conserved, the decrease in mechanical energy in the system is balanced by an increase of heat energy. This heat is energy transferred from motion (kinetic energy) through friction. A swinging pendulum, for example, has alternating kinetic and potential energy, but, over time, this mechanical energy is lost to heat energy by way of friction.

If the whole system is thermally and mechanically isolated (i.e., dU = 0), eventually all the mechanical energy of the pendulum will become heat energy. As a side note, it is this phenomenon that, in the context of cosmology, is referred to as "heat death." Some physicists believe that all the "useful" energy of the universe will eventually become heat energy, resulting in a "winding down" of the universe.

In the case of optical systems, the energy contained in incident light can be lost to heat through absorption. On a microscopic level, as described by quantum mechanics, incident photons can be absorbed by a material, causing an excitation of a particular atom or number of atoms. Although these excited atoms may relax to lower energy states through the emission of photons (thus maintaining the energy in the form of light), the relaxation process may also occur through emission of a phonon, or vibrational mode. These phonons are responsible for heat conduction in a material.

For acoustics, sound waves can lose energy to heating as well. As variations in pressure occur, energy in acoustic vibrations can be transferred to heat energy in the material. Again, the total energy of the system must remain constant, and therefore the sum of the mechanical and thermal energies cannot change as long as the system remains closed.

Skill 12.4   Use the principle of entropy to analyze the operation of heat engines

To understand the second law of thermodynamics, we must first understand the concept of entropy. Entropy is the transformation of energy to a more disordered state and is the measure of how much energy or heat is available for work. The greater the entropy of a system, the less energy is available for work.

A quantitative measure of entropy S is given by the statement that the change in entropy of a system that goes from one state to another in an isothermal and reversible process is the amount of heat absorbed in the process divided by the absolute temperature at which the process occurs.

$$\Delta S = \frac{\Delta Q}{T}$$

Stated more generally, the entropy change that occurs in a state change between two equilibrium states A and B via a reversible process is given by

$$\Delta S_{A \to B} = \int_{A}^{B} \frac{dQ}{T}$$

<u>Problem:</u>   What is the change in entropy of a cube of ice of mass 30g which melts at temperature 0C? The latent heat of fusion of ice is 334 KJ/Kg.

<u>Solution:</u> The amount of heat absorbed by the ice cube =
$30 \times 10^{-3} \times 334 KJ = 10020 J$.

Thus change in entropy = (10020/273)J/K = 36.7 J/K

The second law of thermodynamics may also be stated in the following ways:
1. No machine is 100% efficient.
2. Heat cannot spontaneously pass from a colder to a hotter object.

If we consider a heat engine that absorbs heat Q_h from a hot reservoir at temperature T_h and does work W while rejecting heat Q_c to a cold reservoir at a lower temperature T_c, $Q_h - Q_c$ = W. The efficiency of the engine is the ratio of the work done to the heat absorbed and is given by

$$\varepsilon = \frac{W}{Q_h} = \frac{Q_h - Q_c}{Q_h} = 1 - \frac{Q_c}{Q_h}$$

It is impossible to build a heat engine with 100% efficiency, i.e. one where $Q_c = 0$.

Carnot described an ideal reversible engine, the Carnot engine, that works between two heat reservoirs in a cycle known as the Carnot cycle which consists of two isothermal (12 and 34) and two adiabatic processes (23 and 41) as shown in the diagram below.

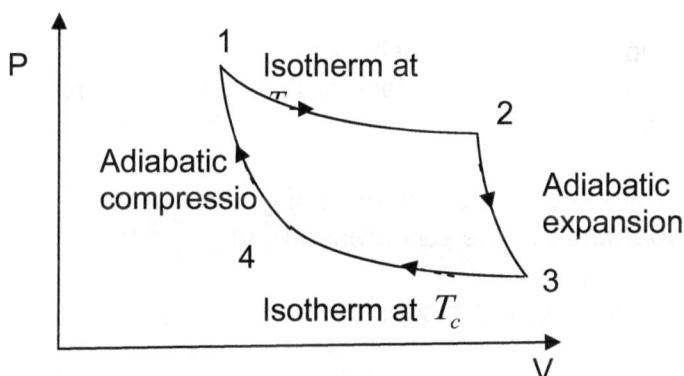

The efficiency of a Carnot engine is given by $\varepsilon = 1 - \dfrac{T_c}{T_h}$ where the temperature values are absolute temperatures. This is the highest efficiency that any engine working between T_c and T_h can reach.

According to Carnot's theorem, no engine working between two heat reservoirs can be more efficient than a reversible engine. All such reversible engines have the same efficiency.

COMPETENCY 13.0   Understand the kinetic-molecular theory and its relationship to thermodynamics

Skill 13.1   Analyze the behavior of an ideal gas in terms of the kinetic-molecular theory

In the case of an ideal gas, which obeys the ideal gas law PV=nRT, work done on or by the system result in a transfer of energy available to do work into heat energy. A particular case would be a confined ideal gas; if the volume V of the gas is decreased, meaning work is done on the system, either the pressure P of the gas increases (in which case the gas acquires energy available to do mechanical work) or the heat energy of the gas (as measured by the temperature T) increases. Alternatively, some combination of changes in these properties takes place. If the gas is allowed to cool, venting heat energy out of the gas, then the total energy of the gas decreases, meaning that, in this case, less energy is available for the gas to do work.

The pressure, temperature and volume relationships for an ideal gas (a gas described by the assumptions of the kinetic molecular theory listed above) are given by the following gas laws:

Boyle's law states that the volume of a fixed amount of gas at constant temperature is inversely proportional to the gas pressure, or:

$$V \propto \frac{1}{P}. \text{ or PV = constant}$$

<u>Problem:</u> A 15 liter container contains gas at a pressure of 5 atmospheres. If the volume of the container is reduced to 3 liters and the temperature remains unchanged, what will be the pressure in the container?

<u>Solution:</u> Since PV is constant by Boyle's law, the product of the initial pressure and volume is equal to the product of the final pressure and volume, i.e.

$$P_1 V_1 = P_2 V_2$$

Therefore, $P_2 = P_1 V_1 / V_2 = 15 \times 5 / 3 atm. = 25 atm.$

Gay-Lussac's law states that the pressure of a fixed amount of gas in a fixed volume is proportional to absolute temperature, or:

$$P \propto T.$$

Charles' law states that the volume of a fixed amount of gas at constant pressure is directly proportional to absolute temperature, or:

$$V \propto T.$$

<u>Problem</u>: A container has 25 liters of gas at 0^0C. What will the volume of the gas be if the temperature is raised to 100^0C at constant pressure?

<u>Solution:</u> From Charles' law we know that $V_1 / T_1 = V_2 / T_2$.

Thus $V_2 = V_1T_2 / T_1 = 25 \times 373 / 273 = 34.2 liters$.

The **combined gas law** uses the above laws to determine a proportionality expression that is used for a constant quantity of gas:

$$V \propto \frac{T}{P}.$$

The combined gas law is often expressed as an equality between identical amounts of an ideal gas at two different states ($n_1=n_2$):

$$\frac{P_1V_1}{T_1} = \frac{P_2V_2}{T_2}.$$

Avogadro's hypothesis states that equal volumes of different gases at the same temperature and pressure contain equal numbers of molecules.
Avogadro's law states that the volume of a gas at constant temperature and pressure is directly proportional to the quantity of gas, or $V \propto n$ where n is the number of moles of gas.

Avogadro's law and the combined gas law yield $V \propto \frac{nT}{P}$. The proportionality constant R--the ideal gas constant--is used to express this proportionality as the ideal gas law:

$$PV = nRT$$

The ideal gas law is useful because it contains all the information of Charles's, Avogadro's, Boyle's, and the combined gas laws in a single expression.

Skill 13.2 Analyze the properties of materials in terms of molecular arrangement and forces

The phase or state of matter (solid, liquid, or gas) is identified by its shape and volume. A solid has a definite shape and volume. A liquid has a definite volume, but no shape. A gas has no shape or volume because it will spread out to occupy the entire space of whatever container it is in. While plasma is really a type of gas, its properties are so unique that it is considered a unique phase of matter. Plasma is a gas that has been ionized; meaning that at least on electron has been removed from some of its atoms. Plasma shares some characteristics with gas, specifically, the high kinetic energy of its molecules.

Thus, plasma exists as a diffuse "cloud," though it sometimes includes tiny grains (this is termed dusty plasma). What most distinguishes plasma from gas is that it is electrically conductive and exhibits a strong response to electromagnetic fields. This property is a consequence of the charged particles that result from the removal of electrons from the molecules in the plasma.

Molecules have kinetic energy (they move around), and they also have intermolecular attractive forces (they stick to each other). The relationship between these two determines whether a collection of molecules will be a gas, liquid, or solid.

A gas has an indefinite shape and an indefinite volume. The kinetic model for a gas is a collection of widely separated molecules, each moving in a random and free fashion, with negligible attractive or repulsive forces between them. Gases will expand to occupy a larger container so there is more space between the molecules. Gases can also be compressed to fit into a small container so the molecules are less separated. Diffusion occurs when one material spreads into or through another. Gases diffuse rapidly and move from one place to another.

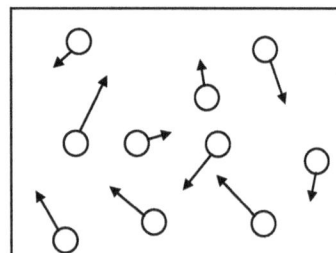

A liquid assumes the shape of the portion of any container that it occupies and has a specific volume. The kinetic model for a liquid is a collection of molecules attracted to each other with sufficient strength to keep them close to each other but with insufficient strength to prevent them from moving around randomly. Liquids have a higher density and are much less compressible than gases because the molecules in a liquid are closer together.

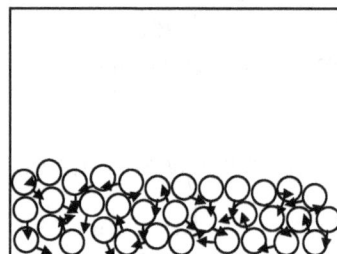

Diffusion occurs more slowly in liquids than in gases because the molecules in a liquid stick to each other and are not completely free to move.

A solid has a definite volume and definite shape. The kinetic model for a solid is a collection of molecules attracted to each other with sufficient strength to essentially lock them in place. Each molecule may vibrate, but it has an average position relative to its neighbors. If these positions form an ordered pattern, the solid is called crystalline. Otherwise, it is called amorphous. Solids have a high density and are almost incompressible because the molecules are close together. Diffusion occurs extremely slowly because the molecules almost never alter their position.

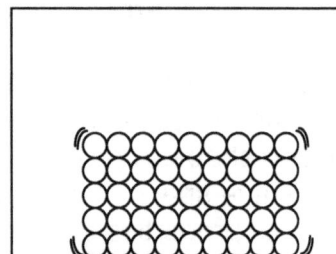

In a gas, the energy of intermolecular forces is much weaker than the kinetic energy of the molecules. Kinetic molecular theory is usually applied to gases and is best applied by imagining ourselves shrinking down to become a molecule and picturing what happens when we bump into other molecules and into container walls.

Gas pressure results from molecular collisions with container walls. The number of molecules striking an area on the walls and the average kinetic energy per molecule are the only factors that contribute to pressure. A higher temperature increases speed and kinetic energy. There are more collisions at higher temperatures, but the average distance between molecules does not change, and thus density does not change in a sealed container. Kinetic molecular theory explains why the pressure and temperature of gases behave the way they do by making a few assumptions, namely:

1) The energies of intermolecular attractive and repulsive forces may be neglected.
2) The average kinetic energy of the molecules is proportional to absolute temperature.
3) Energy can be transferred between molecules during collisions and the collisions are elastic, so the average kinetic energy of the molecules doesn't change due to collisions.
4) The volume of all molecules in a gas is negligible compared to the total volume of the container.

Strictly speaking, molecules also contain some kinetic energy by rotating or experiencing other motions. The motion of a molecule from one place to another is called translation. Translational kinetic energy is the form that is transferred by collisions, and kinetic molecular theory ignores other forms of kinetic energy because they are not proportional to temperature. The following table summarizes the application of kinetic molecular theory to an increase in container volume, number of molecules, and temperature:

Effect of an **increase** in one variable with other two constant	Impact on gas: − = decrease, **0** = no change, **+** = increase						
	Average distance between molecules	Density in a sealed container	Average speed of molecules	Average translational kinetic energy	Collisions with walls per sec.	Collisions per unit area of wall per sec.	Pressure (P)
Vol. of container (V)	+	−	0	0	−	−	−
Number of molecules	−	+	0	0	+	+	+
Temp. (T)	0	0	+	+	+	+	+

Additional details on the kinetic molecular theory may be found at the address below:

http://hyperphysics.phy-astr.gsu.edu/hbase/kinetic/ktcon.html.

An animation of gas particles colliding may be found at the address below:

http://comp.uark.edu/~jgeabana/mol_dyn/

Skill 13.3   Analyze phase changes in terms of kinetic-molecular theory and molecular structure

Phase changes occur when the relative importance of kinetic energy and intermolecular forces is altered sufficiently for a substance to change its state. The transition from gas to liquid is called condensation and from liquid to gas is called vaporization. The transition from liquid to solid is called freezing and from solid to liquid is called melting. The transition from gas to solid is called deposition and from solid to gas is called sublimation.

The relationship between kinetic energy and intermolecular forces determines whether a collection of molecules will be a gas, liquid, or solid. When a material is heated and experiences a phase change from a solid to liquid or liquid to gas, thermal energy is used to break the intermolecular bonds holding the material together. Similarly, bonds are formed with the release of thermal energy when a material changes its phase during cooling.

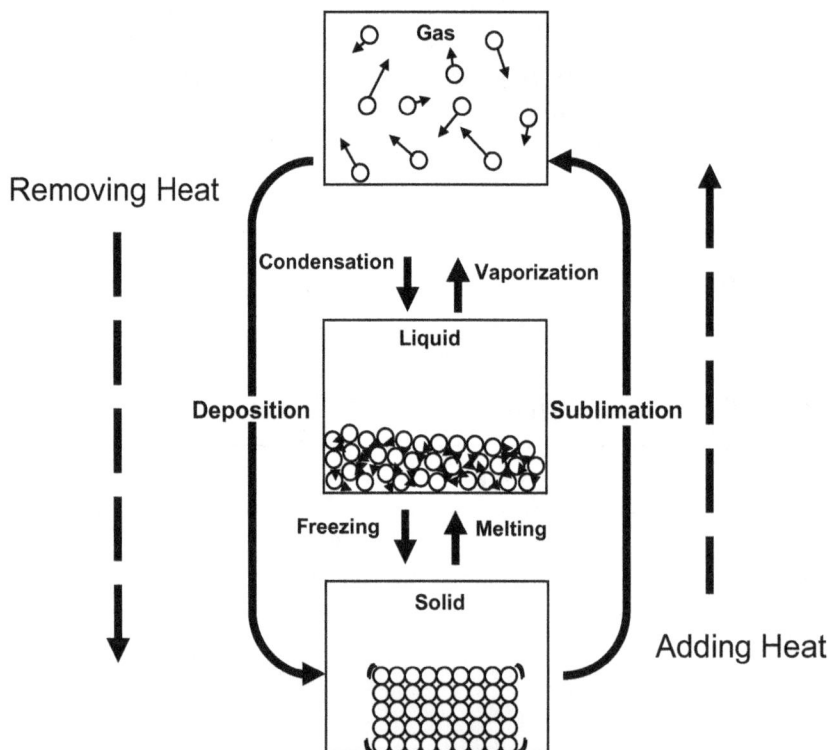

Heat removed from a substance during condensation, freezing, or deposition permits new intermolecular bonds to form, and heat added to a substance during vaporization, melting, or sublimation breaks intermolecular bonds.

During these phase transitions, this latent heat is removed or added with no change in the temperature of the substance because the heat is not being used to alter the speed of the molecules or the kinetic energy when they strike each other or the container walls. Latent heat alters intermolecular bonds.

Skill 13.4   Understand the molecular interpretation of entropy

The simplest statement of the second law of thermodynamics is that the entropy of an isolated system not in equilibrium tends to increase over time. The entropy approaches a maximum value at equilibrium. Below are several common examples in which we see the manifestation of the second law.

- The diffusion of molecules of perfume out of an open bottle
- Even the most carefully designed engine releases some heat and cannot convert all the chemical energy in the fuel into mechanical energy
- A block sliding on a rough surface slows down
- An ice cube sitting on a hot sidewalk melts into a little puddle; we must provide energy to a freezer to facilitate the creation of ice

When discussing the second law, scientists often refer to the "arrow of time". This is to help us conceptualize how the second law forces events to proceed in a certain direction. To understand the direction of the arrow of time, consider some of the examples above; we would never think of them as proceeding in reverse. That is, as time progresses, we would never see a puddle in the hot sun spontaneously freeze into an ice cube or the molecules of perfume dispersed in a room spontaneously re-concentrate themselves in the bottle. The above-mentioned examples are spontaneous as well as irreversible, both characteristic of increased entropy. Entropy change is zero for a complete cycle in a reversible process, a process where infinitesimal quasi-static changes in the absence of dissipative forces can bring a system back to its original state without a net change to the system or its surroundings. All real processes are irreversible. The idea of a reversible process, however, is a useful abstraction that can be a good approximation in some cases.

SUBAREA III.   **ELECTRICITY AND MAGNETISM**

COMPETENCY 14.0   **Understand electric charge, electric fields, electric potential, and capacitance**

Skill 14.1   **Analyze the behavior of an electroscope in given situations**

An electroscope is, fundamentally, a charge-sensing device. Although it involves no digital output, it can be a helpful heuristic tool for understanding some basic concepts surrounding positive and negative charges. A simple design for an electroscope involves a glass container with a metal ball outside and a metal contact inside. Connected to the contact is a thin metal leaf that can be deflected by only a small applied force.

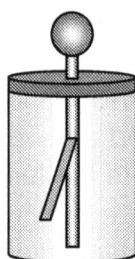

The behavior of the metal leaf inside the electroscope provides insight into the presence of charges in the surrounding environment or in the metal portions of the electroscope itself.

If a charged object is brought near (but does not touch) the electroscope (i.e., the metal ball), the free charge carriers in the metal rearrange themselves to minimize the energy of the system. If the object is negatively charged due to an excess of electrons, for example, the otherwise evenly distributed electrons in the metal of the electroscope will reorient such that a net positive charge is maintained near the charged object. This reorientation leads to a net negative charge in the other portion of the electroscope (i.e., inside the glass container), since the electroscope is initially uncharged. The local net charge results in electrical repulsion between the metal contact and the metal leaf. The leaf is then deflected.

If, while the charged object is near, the metal ball is touched by an observer, for example, the electroscope becomes "grounded," and some of the excess charge in certain areas is discharged. This results in the leaf returning to its original undeflected position. When the charged object is removed (after the observer also breaks contact), the result is that the metal leaf remains deflected. This phenomenon results from the excess charge obtained from grounding the electroscope during the time it was in the presence of a charged object. In the case of the above example, an excess of positive charge would remain and would disperse evenly throughout the metal (i.e., evenly very near the surface).

If the electroscope is grounded once more, with the charged object no longer present, the excess charge will be lost and the leaf will return to its undeflected position.

In addition to the above method of charging the electroscope by induction, it may also be charged by rubbing the outside metal with a charged object, such as a glass rod. The conducting metal allows the excess charge on the rod to further distribute itself into a preferred lower-energy situation. This results in a deflection of the leaf that, as with the previous case, can be reversed by grounding the electroscope.

Skill 14.2   Apply Coulomb's law to determine the forces between charges

Any point charge may experience force resulting from attraction to or repulsion from another charged object. The easiest way to begin analyzing this phenomenon and calculating this force is by considering two point charges. Let us say that the charge on the first point is Q_1, the charge on the second point is Q_2, and the distance between them is r. Their interaction is governed by Coulomb's Law which gives the formula for the force F as:

$$F = k\frac{Q_1 Q_2}{r^2}$$

where $k= 9.0 \times 10^9 \, \frac{N \cdot m^2}{C^2}$ (known as Coulomb's constant)

The charge is a scalar quantity, however, the force has direction. For two point charges, the direction of the force is along a line joining the two charges. Note that the force will be repulsive if the two charges are both positive or both negative and attractive if one charge is positive and the other negative. Thus, a negative force indicates an attractive force.

When more than one point charge is exerting force on a point charge, we simply apply Coulomb's Law multiple times and then combine the forces as we would in any statics problem. Let's examine the process in the following example problem.

Problem: Three point charges are located at the vertices of a right triangle as shown below. Charges, angles, and distances are provided (drawing not to scale). Find the force exerted on the point charge A.

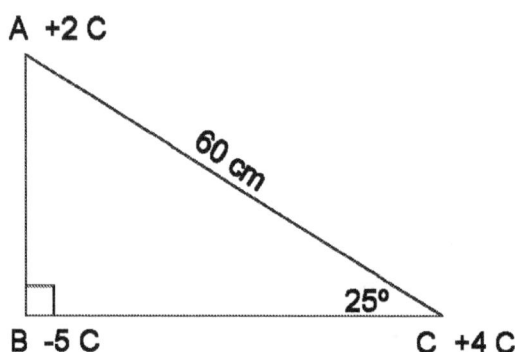

A +2 C

60 cm

25°

B -5 C   C +4 C

Solution: First we find the individual forces exerted on A by point B and point C. We have the information we need to find the magnitude of the force exerted on A by C.

$$F_{AC} = k\frac{Q_1 Q_2}{r^2} = 9 \times 10^9 \frac{N \cdot m^2}{C^2}\left(\frac{4C \times 2C}{(0.6m)^2}\right) = 2 \times 10^{11} N$$

To determine the magnitude of the force exerted on A by B, we must first determine the distance between them.

$$\sin 25° = \frac{r_{AB}}{60cm}$$

$$r_{AB} = 60cm \times \sin 25° = 25cm$$

Now we can determine the force.

$$F_{AB} = k\frac{Q_1 Q_2}{r^2} = 9 \times 10^9 \frac{N \cdot m^2}{C^2}\left(\frac{-5C \times 2C}{(0.25m)^2}\right) = -1.4 \times 10^{12} N$$

We can see that there is an attraction in the direction of B (negative force) and repulsion in the direction of C (positive force). To find the net force, we must consider the direction of these forces (along the line connecting any two point charges). We add them together using the law of cosines.

$$F_A{}^2 = F_{AB}{}^2 + F_{AC}{}^2 - 2F_{AB}F_{AC}\cos 75°$$
$$F_A{}^2 = (-1.4 \times 10^{12}\,N)^2 + (2 \times 10^{11}\,N)^2 - 2(-1.4 \times 10^{12}\,N)(2 \times 10^{11}\,N)^2 \cos 75°$$
$$F_A = 1.5 \times 10^{12}\,N$$

This gives us the magnitude of the net force, now we will find its direction using the law of sines.

$$\frac{\sin \theta}{F_{AC}} = \frac{\sin 75°}{F_A}$$
$$\sin \theta = F_{AC}\frac{\sin 75°}{F_A} = 2 \times 10^{11}\,N\,\frac{\sin 75°}{1.5 \times 10^{12}\,N}$$
$$\theta = 7.3°$$

Thus, the net force on A is 7.3° west of south and has magnitude 1.5 x 10^{12}N. Looking back at our diagram, this makes sense, because A should be attracted to B (pulled straight south) but the repulsion away from C "pushes" this force in a westward direction.

Skill 14.3   Apply principles of electrostatics to determine electric field intensity and electric potential for a given charge distribution

Electric fields can be generated by a single point charge or by a collection of charges in close proximity. The electric field generated from a point charge is given by:

$$E = \frac{kQ}{r^2}$$

where E= the electric field

$k = 9.0 \times 10^9\,\dfrac{N \cdot m^2}{C^2}$ (Coulomb's constant)

Q= the point charge
r= distance from the charge

Electric fields are visualized with field lines, which demonstrate the strength and direction of an electric field. The electric field around a positive charge points away from the charge and the electric field around a negative charge points toward the charge.

While it's easy enough to calculate and visualize the field generated by a single point charge, we can also determine the nature of an electric field produced by a collection of charge simply by adding the vectors from the individual charges. This is known as the superposition principle. The following equation demonstrates how this principle can be used to determine the field resulting from hundreds or thousands of charges.

$$\vec{E}_{total} = \sum_i \vec{E}_i = \vec{E}_1 + \vec{E}_2 + \vec{E}_3 \ldots$$

Electric potential is simply the potential energy per unit of charge. Given this definition, it is clear that electric potential must be measured in joules per coulomb and this unit is known as a volt (J/C=V).

Within an electric field there are typically differences in potential energy. This potential difference may be referred to as voltage. The difference in electrical potential between two points is the amount of work needed to move a unit charge from the first point to the second point.

Stated mathematically, this is: $$V = \frac{W}{Q}$$

where V= the potential difference
W= the work done to move the charge
Q= the charge

We know from mechanics, however, that work is simply force applied over a certain distance. We can combine this with Coulomb's law to find the work done between two charges distance r apart.

$$W = F.r = k\frac{Q_1 Q_2}{r^2}.r = k\frac{Q_1 Q_2}{r}$$

Now we can simply substitute this back into the equation above for electric potential:

$$V_2 = \frac{W}{Q_2} = \frac{k\frac{Q_1 Q_2}{r}}{Q_2} = k\frac{Q_1}{r}$$

Let's examine a sample problem involving electrical potential.

<u>Problem</u>: What is the electric potential at point A due to the 2 shown charges? If a charge of +2.0 C were infinitely far away, how much work would be required to bring it to point A?

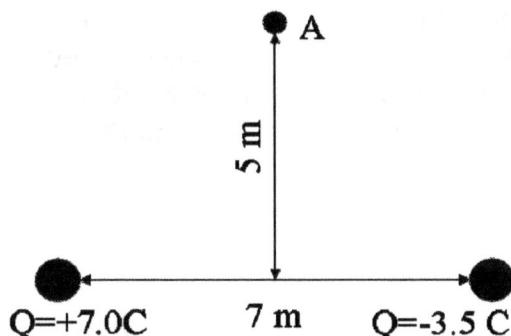

<u>Solution</u>: To determine the electric potential at point A, we simple find and add the potential from the two charges (this is the principle of superposition). From the diagram, we can assume that A is equidistant from each charge. Using the Pythagorean theorem, we determine this distance to be 6.1 m.

$$V = \frac{kq}{r} = k\left(\frac{7.0C}{6.1m} + \frac{-3.5C}{6.1m}\right) = 9 \times 10^9 \, \frac{N.m^2}{C^2}\left(0.57\frac{C}{m}\right) = 5.13 \times 10^9 V$$

Now, let's consider bringing the charged particle to point A. We assume that electric potential of these particle is initially zero because it is infinitely far away. Since now know the potential at point A, we can calculate the work necessary to bring the particle from V=0, i.e. the potential energy of the charge in the electrical field:

$$W = VQ = (5.13 \times 10^9) \times 2J = 10.26 \times 10^9 J$$

The large results for potential and work make it apparent how large the unit coulomb is. For this reason, most problems deal in microcoulombs (μC).

Skill 14.4   Understand the relationship among capacitance, charge, and potential difference

Capacitance (C) is a measure of the stored electric charge per unit electric potential. The mathematical definition is:

$$C = \frac{Q}{V}$$

It follows from the definition above that the units of capacitance are coulombs per volt, a unit known as a farad ($F=C/V$). In circuits, devices called parallel plate capacitors are formed by two closely spaced conductors. The function of capacitors is to store electrical energy. When a voltage is applied, electrical charges build up in both the conductors (typically referred to as plates). These charges on the two plates have equal magnitude but opposite sign. The capacitance of a capacitor is a function of the distance d between the two plates and the area A of the plates:

$$C \approx \frac{\varepsilon A}{d} ; A \gg d^2$$

Capacitance also depends on the permittivity of the non-conducting matter between the plates of the capacitor. This matter may be only air or almost any other non-conducting material and is referred to as a **dielectric**. The permittivity of empty space ε_0 is roughly equivalent to that for air, $\varepsilon_{air} = 8.854 \times 10^{-12}$ C ^{2}/N•m^2. For other materials, the dielectric constant, κ, is the permittivity of the material in relation to air ($\kappa = \varepsilon / \varepsilon_{air}$). The make-up of the dielectric is critical to the capacitor's function because it determines the maximum energy that can be stored by the capacitor. This is because an overly strong electric field will eventually destroy the dielectric.

In summary, a capacitor is "charged" as electrical energy is delivered to it and opposite charges accumulate on the two plates. The two plates generate electric fields and a voltage develops across the dielectric. The energy stored in the capacitor, then, is equal to the amount of work necessary to create this voltage. The mathematical statement of this is:

$$E_{stored} = \frac{1}{2}CV^2 = \frac{1}{2}\frac{Q^2}{C} = \frac{1}{2}VQ$$

The work per unit volume or the **electric field energy density** within a capacitor can be shown to be $\eta = \frac{1}{2}\varepsilon E^2$. This result is generally valid for the energy per unit volume of any electrostatic field, not only for a constant field within a capacitor.

Problem: Imagine that a parallel plate capacitor has an area of 10.00 cm^2 and a capacitance of 4.50 pF. The capacitor is connected to a 12.0 V battery. The capacitor is completely charged and then the battery is removed. What is the separation of the plates in the capacitor? How much energy is stored between the plates? We've assumed that this capacitor initially had no dielectric (i.e., only air between the plates) but now imagine it has a Mylar dielectric that fully fills the space. What will the new capacitance be? (for Mylar, □=3.5)

Solution: To determine the separation of the plates, we use our equation for a capacitor:

$$C = \frac{\varepsilon_0 A}{d}$$

We can simply solve for d and plug in our values:

$$d = \varepsilon_0 \frac{A}{C} = \left(8.854 \times 10^{-12} \frac{C}{N \cdot m^2}\right) \frac{10 \times 10^{-4} m^2}{4.5 \times 10^{-12} F} = 1.97 \times 10^{-3} m = 1.97 mm$$

Similarly, to find stored energy, we simply employ the equation above:

$$E_{stored} = \frac{1}{2} QV$$

But we don't yet know the charge Q, so we must first find it from the definition of capacitance:

$$C = \frac{Q}{V}$$

$$Q = CV = (4.5 \times 10^{-12}) \times (12V) = 5.4 \times 10^{-11} C$$

Now we can find the stored energy:

$$E_{stored} = \frac{1}{2} QV = \frac{1}{2}(5.4 \times 10^{-11} C)(12V) = 3.24 \times 10^{-10} J$$

To find the capacitance with a Mylar dielectric, we again use the equation for capacitance of a parallel plate capacitor. Note that the new capacitance can be found by multiplying the original capacitance by κ:

$$C = \frac{\kappa_{Mylar} \varepsilon_0 A}{d} = \kappa_{Mylar} C_0 = 3.5 \times 4.5 pF = 15.75 pF$$

COMPETENCY 15.0   Understand the components and properties of direct current circuits

Skill 15.1   Analyze a DC circuit in terms of conservation of energy and conservation of charge

Within an electric field there are typically differences in potential energy. This potential difference is usually referred to as voltage. Since electric potential is the potential energy per unit of charge, it is clear that electric potential must be measured in joules per coulomb and this unit is known as a volt. Electrical current, however, is the rate at which charge flows past a given point. Current is measured in amperes which is equivalent to coulombs/ second.

When electrical charge flows through a conductor, both current and a potential difference will be present. It is logical that a greater potential difference will lead to a faster flow of charge and, therefore, a higher current:

$$V \alpha I$$

As in any other proportionality relationship, we can simply add a proportionality constant:

$$V = k I$$

In the case of electricity, the proportionality constant is a measure of the material's resistance to current. This quantity is known as resistance and is a constant for a given material at a given temperature regardless of voltage or current. Restating our equation:

$$V = R I$$

This expression is **Ohm's Law**. Note that resistance must have units of $\dfrac{J \cdot s}{C^2}$ and this is known as ohm (Ω).

We can draw a rough analogy between this electrical phenomenon and simple mechanics. Imagine a block sliding down an inclined plane under the force of gravity. The speed of the block as it travels from one point to another is directly proportional to the difference in gravitational potential energy between those points. However, the block's speed can be slowed by the friction between the block and plane. Thus friction is similar to resistance, voltage similar to the difference in gravitational potential energy, and current similar to speed. The voltage drop, like gravity, provides the driving force while resistance impedes the flow.

Kirchoff's Laws are a pair of laws that apply to conservation of charge and energy in circuits and were developed by Gustav Kirchoff.

Kirchoff's Current Law: At any point in a circuit where charge density is constant, the sum of currents flowing toward the point must be equal to the sum of currents flowing away from that point. **Kirchoff's Voltage Law**: The sum of the electrical potential differences around a circuit must be zero. While these statements may seem rather simple, they can be very useful in analyzing DC circuits, those involving constant circuit voltages and currents.

Problem:
The circuit diagram at right shows three resistors connected to a battery in series. A current of 1.0 A is generated by the battery. The potential drop across R_1, R_2, and R_3 are 5V, 6V, and 10V. What is the total voltage supplied by the battery?

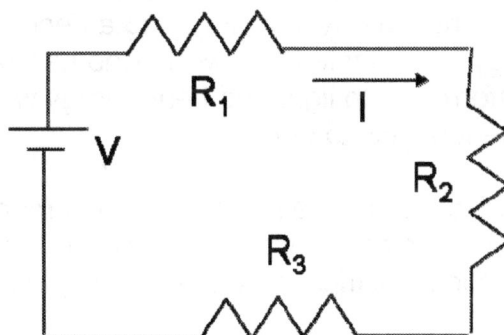

Solution:
Kirchoff's Voltage Law tells us that the total voltage supplied by the battery must be equal to the total voltage drop across the circuit. Therefore:

$$V_{battery} = V_{R_1} + V_{R_2} + V_{R_3} = 5V + 6V + 10V = 21V$$

Problem:
The circuit diagram at right shows three resistors wired in parallel with a 12V battery. The resistances of R_1, R_2, and R_3 are 4 Ω, 5 Ω, and 6 Ω, respectively. What is the total current?

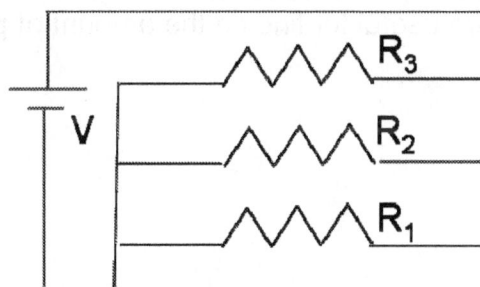

Solution:
This is a more complicated problem. Because the resistors are wired in parallel, we know that the voltage entering each resistor must be the same and equal to that supplied by the battery. We can combine this knowledge with Ohm's Law (see next section) to determine the current across each resistor:

$$I_1 = \frac{V_1}{R_1} = \frac{12V}{4\Omega} = 3A \qquad I_2 = \frac{V_2}{R_2} = \frac{12V}{5\Omega} = 2.4A \qquad I_3 = \frac{V_3}{R_3} = \frac{12V}{6\Omega} = 2A$$

Finally, we use Kirchoff's Current Law to find the total current:

$$I = I_1 + I_2 + I_3 = 3A + 2.4A + 2A = 7.4A$$

Skill 15.2　Analyze energy relationships and transformations in electric circuits

The first law of thermodynamics applies in a number of situations. Ideally, in many simple models of various phenomena, undesirable effects such as friction, resistance and absorption are ignored. In reality, however, these are unavoidable consequences of the nature of the phenomena. In the context of electrical circuits, collisions in wires or devices impedes the flow of electrons and can convert mechanical energy, which originates from an applied electric or magnetic field, into heat energy. If the heat is dissipated from the circuit then that heat loss represents energy loss, implying a decrease in the circuit's efficiency since less energy is available to do work. Another example is electrical energy being transformed into light and heat energy when light bulb is turned on and the filament begins to glow.

Electrical power is a measure of how much energy is expended or how much work can be done by an electrical current and has units of watts (W). To determine electrical power, we simply use Joule's law:

$$P = IV$$

where P=power
I=current
V=voltage

If we combine this with Ohm's law (V=IR), we generate two new equations that are useful for finding the amount of power dissipated by a resistor:

$$P = I^2 R$$

$$P = \frac{V^2}{R}$$

Problem: How much power is dissipated by a 1 kΩ resistor with a 50V voltage drop through it?

Solution:

$$P = \frac{V^2}{R} = \frac{(50V)^2}{1000\Omega} = 2.5W$$

Skill 15.3   Interpret schematic diagrams of electric circuits

The two most important elements in simple circuits are resistors and capacitors. Often resistors and capacitors are used together in series or parallel. Two components are in series if one end of the first element is connected to one end of the second component. The components are in parallel if both ends of one element are connected to the corresponding ends of another. A series circuit has a single path for current flow through all of its elements. A parallel circuit is one that requires more than one path for current flow in order to reach all of the circuit elements.

Below is a diagram demonstrating a simple circuit with resistors in parallel (on right) and in series (on left). Note the symbols used for a battery (noted V) and the resistors (noted R).

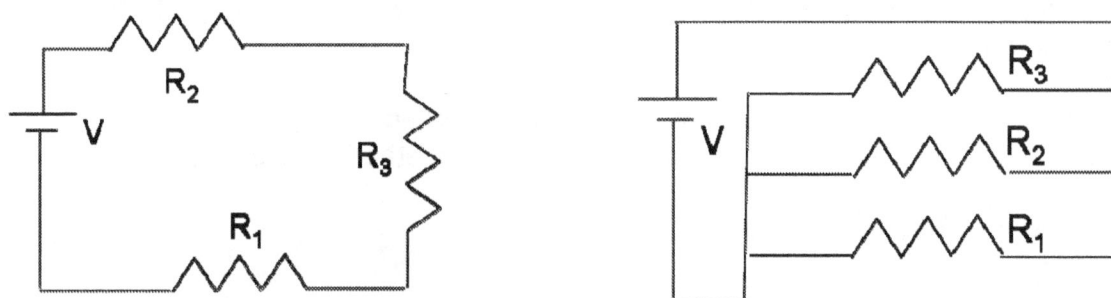

Skill 15.4   Apply principles of DC circuits to reduce a complex circuit to a simpler equivalent circuit

When the resistors are placed in series, the current through each one will be the same. When they are placed in parallel, the voltage through each one will be the same. To understand basic circuitry, it is important to master the rules by which the equivalent resistance (R_{eq}) or capacitance (C_{eq}) can be calculated from a number of resistors or capacitors:

Resistors in parallel:
$$\frac{1}{R_{eq}} = \frac{1}{R_1} + \frac{1}{R_2} + \cdots + \frac{1}{R_n}$$

Resistors in series:
$$R_{eq} = R_1 + R_2 + \cdots + R_n$$

Capacitors in parallel:
$$C_{eq} = C_1 + C_2 + \cdots + C_n$$

Capacitors in series:
$$\frac{1}{C_{eq}} = \frac{1}{C_1} + \frac{1}{C_2} + \cdots + \frac{1}{C_n}$$

COMPETENCY 16.0   Understand magnetic fields and electromagnetic induction

Skill 16.1   Determine the orientation and magnitude of a magnetic field in a given situation

Conductors through which electrical currents travel will produce magnetic fields: The magnetic field *dB* induced at a distance *r* by an element of current *Idl* flowing through a wire element of length *dl* is given by the **Biot-Savart** law

$$dB = \frac{\mu_0}{4\pi} \frac{Idl \times \hat{r}}{r^2}$$

where μ_0 is a constant known as the permeability of free space and $\hat{r}$ is the unit vector pointing from the current element to the point where the magnetic field is calculated.

An alternate statement of this law is **Ampere's law** according to which the line integral of *B.dl* around any closed path enclosing a steady current *I* is given by

$$\oint_C B \cdot dl = \mu_0 I$$

The basis of this phenomenon is the same no matter what the shape of the conductor, but we will consider three common situations:

Straight Wire
Around a current-carrying straight wire, the magnetic field lines form concentric circles around the wire. The direction of the magnetic field is given by the right-hand rule: When the thumb of the right hand points in the direction of the current, the fingers curl around the wire in the direction of the magnetic field. Note the direction of the current and magnetic field in the diagram.

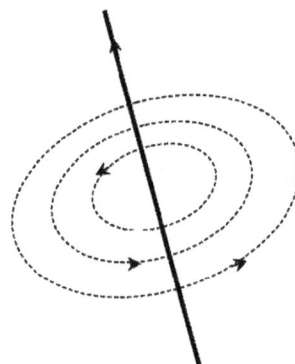

To find the magnetic field of an infinitely long (allowing us to disregarding end effects) we apply Ampere's Law to a circular path at a distance r around the wire:

$$B = \frac{\mu_0 I}{2\pi r}$$

where μ_0=the permeability of free space ($4\pi \times 10^{-7}$ T·m/A)
I=current
r=distance from the wire

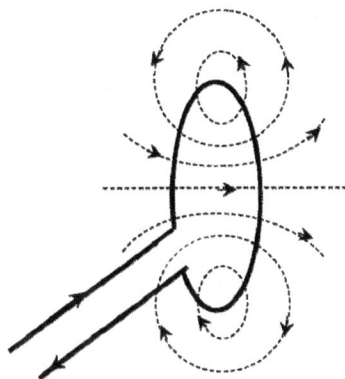

Loops

Like the straight wire from which it's been made, a looped wire has magnetic field lines that form concentric circles with direction following the right-hand rule. However, the field are additive in the center of the loop creating a field like the one shown. The magnetic field of a loop is found similarly to that for a straight wire.

In the center of the loop, the magnetic field is:

$$B = \frac{\mu_0 I}{2r}$$

Solenoids

A solenoid is essentially a coil of conduction wire wrapped around a central object. This means it is a series of loops and the magnetic field is similarly a sum of the fields that would form around several loops, as shown.

The magnetic field of a solenoid can be found as with the following equation:

$$B = \mu_0 n I$$

In this equation, n is turn density, which is simply the number of turns divided by the length of the solenoid.

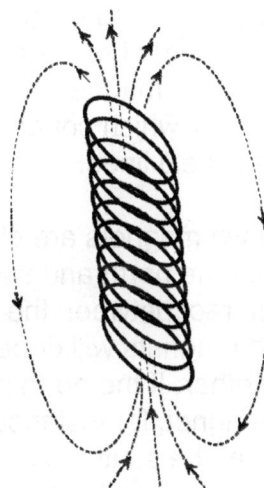

Magnetic field lines are a good way to visualize a magnetic field. The distance between magnetic fields lines indicates the strength of the magnetic field such that the lines are closer together near the poles of the magnets where the magnetic field is the strongest. The lines spread out above and below the middle of the magnet, as the field is weakest at those points furthest from the two poles. The SI unit for magnetic field known as magnetic induction is Tesla(T) given by 1T = 1 N.s/(C.m) = 1 N/(A.m). Magnetic fields are often expressed in the smaller unit Gauss (G) (1 T = 10,000 G). Magnetic field lines always point from the north pole of a magnet to the south pole.

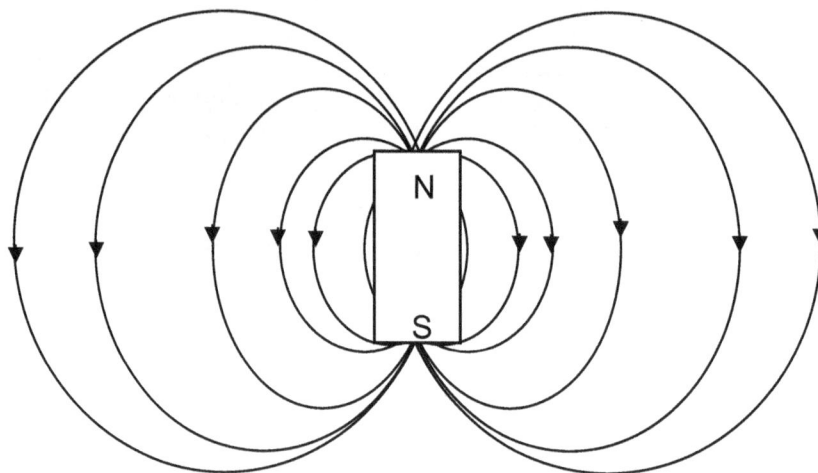

Magnetic field lines can be plotted with a magnetized needle that is free to turn in 3 dimensions. Usually a compass needle is used in demonstrations. The direction tangent to the magnetic field line is the direction the compass needle will point in a magnetic field. Iron filings spread on a flat surface or magnetic field viewing film which contains a slurry of iron filings are another way to see magnetic field lines.

When two magnets are placed close to one another magnetic field lines form between the north and south pole of each magnet individually and the field line also interact between the two magnets. The magnetic field in the area between the two magnets will depend on how the magnets are oriented with respect to one another. If the north pole of one magnet is placed near the south pole of the other magnet, for instance, you can see field lines starting at that north pole and ending at the south pole of the other magnet. If additional magnets and metal objects are placed in the arrangement the interactions become more complicated and various patterns are formed by the magnetic field lines.

Skill 16.2   Determine the magnitude and direction of the force on a charge or charges moving in a magnetic field

The magnetic force exerted on a charge moving in a magnetic field depends on the size and velocity of the charge as well as the magnitude of the magnetic field. One important fact to remember is that only the velocity of the charge in a direction perpendicular to the magnetic field will affect the force exerted. Therefore, a charge moving parallel to the magnetic field will have no force acting upon it whereas a charge will feel the greatest force when moving perpendicular to the magnetic field.

The direction of the magnetic force is always at a right angle to the plane formed by the velocity vector v and the magnetic field B and is given by applying the right hand rule - if the fingers of the right hand are curled in a way that seems to rotate the v vector into the B vector, the thumb points in the direction of the force. The magnitude of the force is equal to the cross product of the velocity of the charge with the magnetic field multiplied by the magnitude of the charge.

$$F = q \ (v \times B) \quad or \quad F = q \ v \ B \sin(\theta)$$

where θ= the angle formed between the vectors of velocity of the charge and direction of magnetic field

Problem: Assuming we have a particle of 1×10^{-6} kg that has a charge of -8 coulombs that is moving perpendicular to a magnetic field in a clockwise direction on a circular path with a radius of 2 m and a speed of 2000 m/s, let's determine the magnitude and direction of the magnetic field acting upon it.

Solution: We know the mass, charge, speed, and path radius of the charged particle. Combining the equation above with the equation for centripetal force we get

$$qvB = \frac{mv^2}{r} \quad or \quad B = \frac{mv}{qr}$$

Thus B= $(1 \times 10^{-6}$ kg) (2000m/s) / (-8 C)(2 m) = 1.25×10^{-4} Tesla

Since the particle is moving in a clockwise direction, we use the right hand rule and point our fingers clockwise along a circular path in the plane of the paper while pointing the thumb towards the center in the direction of the centripetal force. This requires the fingers to curl in a way that indicates that the magnetic field is pointing out of the page. However, since the particle has a negative charge we must reverse the final direction of the magnetic field into the page.

A mass spectrometer measures the mass to charge ratio of ions using a setup similar to the one described above. m/q is determined by measuring the path radius of particles of known velocity moving in a known magnetic field.

Skill 16.3   Analyze factors that affect the magnitude of an induced electromotive force (EMF)

When the magnetic flux through a coil is changed, a voltage is produced which is known as induced electromagnetic force. Magnetic flux is a term used to describe the number of magnetic fields lines that pass through an area and is described by the equation:

$$\Phi = B\, A\, \cos\theta$$

where Φ = the angle between the magnetic field B, and the normal to the plane of the coil of area A

By changing any of these three inputs, magnetic field, area of coil, or angle between field and coil, the flux will change and an EMF can be induced. The speed at which these changes occur also affects the magnitude of the EMF, as a more rapid transition generates more EMF than a gradual one. This is described by **Faraday's law** of induction:

$$\varepsilon = -N\, \Delta\Phi \,/\, \Delta t$$

where ε = emf induced, N is the number of loops in a coil, t is time, and Φ is magnetic flux

The negative sign signifies **Lenz's law** which states that induced emf in a coil acts to oppose any change in magnetic flux. Thus the current flows in a way that creates a magnetic field in the direction opposing the change in flux. See the next section for the right-hand rule that determines the direction of the induced current.

Consider a coil lying flat on the page with a square cross section that is 10 cm by 5 cm. The coil consists of 10 loops and has a magnetic field of 0.5 T passing through it coming out of the page. Let's find the induced EMF when the magnetic field is changed to 0.8 T in 2 seconds.

First, let's find the initial magnetic flux: Φ_i
$\Phi_i = BA \cos\theta = (.5\ T)\,(.05\ m)\,(.1 m)\cos 0° = 0.0025\ T\ m^2$

And the final magnetic flux: Φ_f
$\Phi f = BA \cos\theta = (0.8\ T)\,(.05\ m)\,(.1 m)\cos 0° = 0.004\ T\ m^2$

The induced emf is calculated then by
$\varepsilon = -N\, \Delta\Phi \,/\, \Delta t = -10\,(.004\ T\ m^2 - .0025\ T\ m^2)\,/\,2\ s = -0.0075$ volts.

To determine the direction the current flows in the coil we need to apply the right hand rule and Lenz's law. The magnetic flux is being increased out of the page, with your thumb pointing up the fingers are coiling counterclockwise. However, Lenz's law tells us the current will oppose the change in flux so the current in the coil will be flowing clockwise.

Skill 16.4 Analyze the use of electromagnetism in technology

Electromagnetism is the foundation for a vast number of modern technologies ranging from computers to communications equipment. More mundane technologies such as motors and generators are also based upon the principles of electromagnetism. The particular understanding of electrodynamics can either be in terms of quantum mechanics (quantum electrodynamics) or classical electrodynamics, depending on the type of phenomenon being analyzed. In classical electrodynamics, which is a sufficient approximation for most situations, the electric field, resulting from electric charge, and the magnetic field, resulting from moving charges, are the parameters of interest and are related through Maxwell's equations.

Motors
Electric motors are found in many common appliances such as fans and washing machines. The operation of a motor is based on the principle that a magnetic field exerts a force on a current carrying conductor. This force is essentially due to the fact that the current carrying conductor itself generates a magnetic field; the basic principle that governs the behavior of an electromagnet. In a motor, this idea is used to convert electrical energy into mechanical energy, most commonly rotational energy. Thus the components of the simplest motors must include a strong magnet and a current-carrying coil placed in the magnetic field in such a way that the force on it causes it to rotate.

A typical motor is composed of a stationary portion, called the stator, and a rotating (or moving) portion, called the rotor. Coils of wire that serve as electromagnets are wound on the armature, which can be either the stator or the rotor, and are powered by an electric source. Motors use electric current to generate a magnetic field around an electromagnet, which results in a rotational force due to the presence of an external magnetic field (either from permanent magnets or electromagnets). The designs of various motors can differ dramatically, but the general principles of electromagnetism that describe their operation are generally the same.

Generators
Generators are in effect "reverse motors". They exploit electromagnetic induction to generate electricity. Thus, if a coil of wire is rotated in a magnetic field, an alternating EMF is produced which allows current to flow. Any number of energy sources can be used to rotate the coil, including combustion, nuclear fission, flowing water or other sources. Therefore, generators are devices that convert mechanical or other forms of energy into electrical energy.

Meters

A number of different types of meters use electromagnetism or are designed to measure certain electromagnetic parameters. For example, older forms of ammeters (galvanometers), when supplied with a current, provided a measurement through the deflection of a spring-loaded needle. A coil connected to the needle acted as an electromagnet which, in the presence of a permanent magnetic field, would be deflected in the same manner as a rotor, as mentioned previously. The strength of the electromagnet, and thus the extent of the deflection, is proportional to the current. Further, the spring limits the deflection in such a manner that a reasonably accurate measurement of the current is provided.

Magnetic Media

Although the cassette tape has fallen out of favor with popular culture, magnetic media are still widely used for information storage. Magnetic strips on the back of credit and identification cards, computer hard drive disks and magnetic tapes (such as those contained in cassettes) are all examples of magnetic media.

The principle underlying magnetic media is the sequential magnetization of a region of the medium. A special head is able to detect spatial magnetic fluctuations in the medium which are then converted into an electrical signal. The head is often able to "write" to the medium as well. The electrical signal from the medium can be converted into sound or video, as with the video or audio cassette player, or it can be digitized for use in a computer.

COMPETENCY 17.0   Understand alternating currents and the operation of conductors, semiconductors, and superconductors

Skill 17.1   Analyze the current, voltage, and phase relationships in an RLC circuit

Within an electric field there are typically differences in potential energy. This potential difference is also observed in conductors, such as the wiring in circuits. This potential difference is usually referred to as voltage. Since electric potential is the potential energy per unit of charge, it is clear that electric potential must be measured in joules per coulomb and this unit is known as a volt. Electrical current, however, is the rate at which charge flows past a given point in the circuit. Current is measured in amperes which is equivalent to coulombs/second.

When electrical charge flows through a circuit, both current and a potential difference will be present. It is logical that a greater potential difference will lead to a faster flow of charge and, therefore, a higher current:

$$V \alpha I$$

As in any other proportional relationships, we can simply add a proportionality constant:

$$V = kI$$

In the case of electricity, the proportionality constant is a measure of the material's resistance to current. This quantity is known as resistance and is a constant for a given material at a given temperature regardless of voltage or current. Restating our equation:

$$V = RI$$

This expression is Ohm's Law. Note that resistance must have units of $\dfrac{J \cdot s}{C^2}$ and this is known as ohm (Ω).

We can draw a rough analogy between this electrical phenomena and simple mechanics. Imagine a block sliding down an incline plane under the force of gravity. The speed of the block as it travels from one point to another is directly proportional to the difference in gravitational potential energy between those points. However, the block's speed can be slowed by the friction between the block and plane. Thus friction is similar to resistance, voltage similar to the difference in gravitational potential energy, and current similar to speed. The voltage drop, like gravity, provides the driving force while resistance impedes the flow.

Skill 17.2 Describe the flow of electrons in conductors and semiconductors

Semiconductors, conductors and superconductors are differentiated by how "easily" current can flow in the presence of an applied electric field. Dielectrics (insulators) conduct little or no current in the presence of an applied electric field, as the charged components of the material (for example, atoms and their associated electrons) are tightly bound and are not free to move within the material. In the case of materials with some amount of conductivity, so-called valence (higher energy level) electrons are only loosely bound and may move among positive charge centers (i.e., atoms or ions).

Conductors, such as metals like aluminum and iron, have numerous valence electrons distributed among the atoms that compose the material. These electrons can move freely, especially under the influence of an applied electric field. In the absence of any applied field, these electrons move randomly, resulting in no net current. If a static electric field is applied, the mobile electrons reorient themselves to minimize the energy of the system and create an equipotential on the surface of the metal (in the ideal case of infinite conductivity).The highly mobile charge carriers do not allow a net field inside the metal; thus conductors can "shield" electromagnetic fields.

Semiconductors, such as silicon and germanium, are materials that can neither be described as conductors, nor as insulators. A certain number of charge carriers are mobile in the semiconductor, but this number is nowhere near the free charge populations of conductors such as metals. "Doping" of a semiconductor by adding so-called donor atoms or acceptor atoms to the intrinsic (or pure) semiconductor can increase the conductivity of the material. Furthermore, combination of a number of differently doped semiconductors (such as donor-doped silicon and acceptor-doped silicon) can produce a device with beneficial electrical characteristics (such as the diode). The conductivity of such devices can be controlled by applying voltages across specific portions of the device.

A useful way to visualize the relationship of conductors and semiconductors is by way of energy band diagrams. For purposes of comparison, an example of an insulator is included here as well.

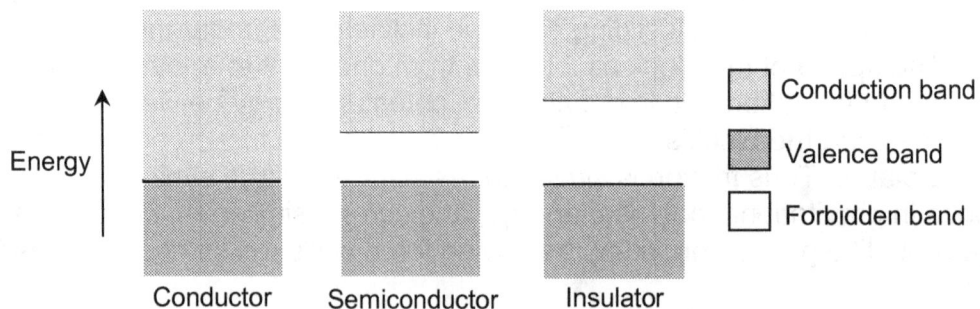

Skill 17.3   Analyze the function of a solid-state device in a given electrical circuit

Prior to the 1960s most electronic apparatuses relied on either vacuum tubes or mechanical devices, typically relays. As semiconductors became increasingly available, however, vacuum tubes were replaced with solid state devices. Vacuum tubes conduct electrons through a heated vacuum but semiconductors allow electrons to flow through them while still solid. Solid state has come to mean any circuit that does not contain the aforementioned vacuum tubes and, in short, operates with "no moving parts". The lack of these elements make solid state devices more resistant to physical stressors, such as vibration, and more durable in general, since they are less susceptible to wear. A familiar example of this are the solid state "flash cards" that are popular for data storage. Previously, hard disk players were primarily used for a similar function but their moving parts make them of limited durability and less practical to transport.

Semiconductors are extremely useful because their conductive properties can be controlled. This is done by "doping" or introducing impurities. The impurity or dopant is used to introduce extra electrons or extra free orbital space that can be filled with electrons. This allows much freer movement of electron and, therefore, flow of current through the material. Semiconductors doped to contain extra electrons are known as N-type, while those doped to contain extra "holes" are known a P-type. Typically, either type of semiconductor can be made from the same base material. For instance, silicon can be doped with boron to create a P-type semiconductor or with phosphorus to create an N-type semiconductor. Two of the most common and important semiconductor devices today are diodes and transistors.

Diodes: Diodes restrict the direction of current (electron) flow in a given direction. Somewhat analagous to a check valve, they allow the electrical current to flow in one direction, but not in the other. Thus diodes are found in almost all circuits where single direction current flow is required. While early diodes were made from vacuum tubes, today most are made from semiconductors. This is often done by desiging a diode with P-type semiconduting material on one side and N-type on the other. The current flows into the P-type side and on to the N-type side, but cannot flow in the opposite direction. Alternatively, the diode can be made from a conducting metal and a semiconductor and function in a similar fashion.

<u>Transistors:</u> Just like the vaccum tubes they replaced, transistors control current flow. They serve a variety of functions in circtuits and can act as amplifiers, switches, voltage regulators, signal modulators, or oscillators. They are perhaps the most important building block of modern circuitry and electronics. As in solid state diodes, the properties of semiconductors are exploited in transistors to control the flow of current. However, where the diode is analagous to a check valve, a transistor function more like a tap on a sink and is able to control the rate of current flow or eliminate it all together. Today, most transistors are either bipolar junction transistors (BJT) or field effect transistors (FET). Both accomplish control of current flow, but it is done by applying current in the BJTs and voltage in the

Skill 17.4   Recognize the properties of superconductors

According to the theory of energy bands, as derived from quantum mechanics, electrons in the ground state reside in the valence band and are bound to their associated atoms or molecules. If the electrons gain sufficient energy, such as through heat, they can jump across the forbidden band (in which no electrons may exist) to the conduction band, thus becoming free electrons that can form a current in the presence of an applied field. For conductors, the valence band and conduction band meet or overlap, allowing electrons to easily jump to unoccupied conduction states. Thus, conductors have an abundance of free electrons. It is noteworthy that only two bands are shown in the diagram above, but that an infinite number of bands may exist at higher energies. A more general statement of the difference in band structures is that, for insulators and semiconductors, the valence electrons fill up all the states in a particular band, leaving a gap between the highest energy valence electrons and the next available band. The difference between these two types of materials is simply a matter of the "size" of the forbidden band. For conductors, the band that contains the highest energy electrons has additional available states.

A nearly ideal conducting material is a superconductor. As a material increases in temperature, increased vibrational motion of the atoms or molecules leads to decreased charge carrier mobility and decreased conductivity. In the case of semiconductors, the increase in free carrier population outweighs the loss in mobility of the charge carriers, meaning that the semiconductor increases in conductivity as temperature increases. Superconductors, on the other hand, reach their peak conductivity at extremely low temperatures (although there are currently numerous efforts to achieve superconductivity at higher and higher temperatures, with room temperature or higher being the ultimate goal). The critical temperature of the material is the temperature at which superconducting properties emerge. At this temperature, the material has a nearly infinite conductivity and maintains an almost perfect equipotential across its surface when in the presence of a static electric field. Inside a superconductor, the electric field is virtually zero at all times. As a result, the time derivative of the electric flux density is zero, and, by Maxwell's equations, the magnetic flux density must likewise be zero.

$$\nabla \times \mathbf{H} = \frac{\partial \mathbf{D}}{\partial t} + \mathbf{J}$$

Since the electric field is also zero, the current density **J** inside the superconductor must also be zero (or very nearly so). This elimination of the magnetic flux density inside a superconductor is called the Meissner effect.

COMPETENCY 18.0   **Understand waves and wave motion, and analyze problems involving wave motion**

Skill 18.1   **Compare the transfer of energy and momentum in longitudinal and transverse waves**

Waves can be described as a propagating disturbance in a medium or local oscillation of a material. Waves transmit energy, but the medium itself does not propagate. For example, a wave can travel on a slinky, but the slinky as a unit does not move. Since mechanical waves require a medium, they are unlike electromagnetic waves because they can not exist in a vacuum.

Waves may be transverse or longitudinal. Returning to the Slinky example, the individual atoms can be made to oscillate from their equilibrium positions in two distinct directions. In transverse waves, the displacement of the medium's particles is perpendicular to the direction of wave propagation. In longitudinal waves, the displacement of atoms or molecules is in the same direction as the wave's propagation. Sound waves are compression or longitudinal waves.

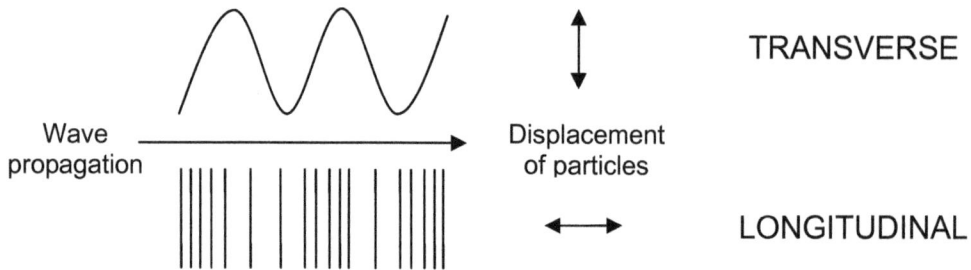

Waves are simply disturbances that propagate through space and time. Most waves must propagate through a medium, which may be solid, liquid, or gas. However, some electromagnetic waves do not require a medium and can move through a vacuum. Most waves transfer energy from their source to their destination, typically without actually transferring molecules of the medium through which they travel. The individual particles of the medium are temporarily displaced but eventually return to their equilibrium positions. As the particles oscillate, they transfer energy and momentum to a neighboring particle which, in turn, moves back and forth around its equilibrium position while transferring energy and momentum to another particle and so on.

All around us, there are examples of waves including ocean waves, sound, radio waves, microwaves, seismic waves, sunlight, x-rays, and radioactive gamma rays. Whether these waves actually displace media or simply carry energy, their positions fluctuate as they move through time and space. Often these fluctuations are regular and we can use both diagrams and mathematical equations to understand the pattern a wave follows in space and how quickly it moves.

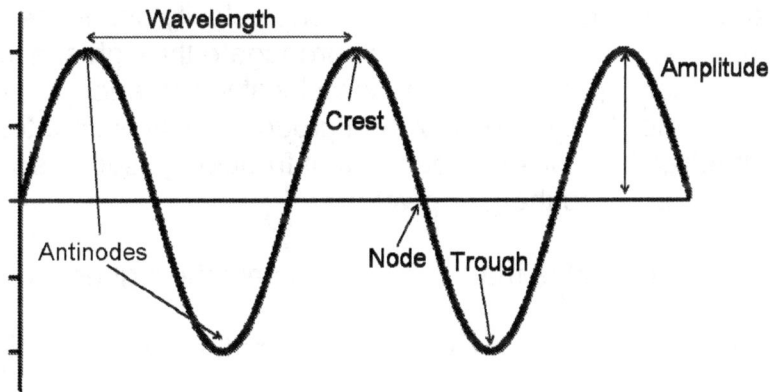

This diagram of a sinusoidal wave shows us displacement caused by the wave as it propagates through a medium. This displacement can be graphed against either time or distance. Note how displacement depends on the distance which the wave has traveled/ how much time has elapsed. So if we chose a particular displacement (let's say the crest), the wave will return to that displacement value (i.e., crest again) after one period (T) or one wavelength (λ).

The general equation for a wave is a partial differential equation which can be simplified to express the behavior of commonly encountered harmonic waveforms. For instance, for a standing wave:

$$y(z,t) = A(z,t)\sin(kz - \omega t + \phi)$$

where y=displacement
z=distance
t=time
k=wave number
ω=angular frequency
ϕ=phase
A(z,t)=the amplitude envelope of the wave

The important concept to note is that y is a function of both z and t. This means that the wave's position depends on both time and distance, just as was seen in the diagram above.

Skill 18.2   Analyze the characteristics of waves

To fully understand waves, it is important to understand many of the terms used to characterize them.

Wave velocity: Two velocities are used to describe waves. The first is phase velocity, which is the rate at which a wave propagates. For instance, if you followed a single crest of a wave, it would appear to move at the phase velocity. The second type of velocity is known as group velocity and is the speed at which variations in the wave's amplitude shape propagate through space. Group velocity is often conceptualized as the velocity at which energy is transmitted by a wave. Phase velocity is denoted v_p and group velocity is denoted v_g. In a medium with refractive index independent of frequency, such as vacuum, the phase velocity is equal to the group velocity.

Crest: The maximum value that a wave assumes; the highest point.

Trough: The lowest value that a wave assumes; the lowest point.

Nodes: The points on a wave with minimal amplitude.

Antinodes: The farthest point from the node on the amplitude axis; both the crests and the troughs are antinodes.

Amplitude: The distance from the wave's highest point (the crest) to the equilibrium point. This is a measure of the maximum disturbance caused by the wave and is typically denoted by A.

Wavelength: The distance between any two sequential troughs or crests denoted λ and representing a complete cycle in the repeated wave pattern.

Period: The time required for a complete wavelength or cycle to pass a given point. The period of a wave is usually denoted T.

Frequency: The number of periods or cycles per unit time (usually a second). The frequency is denoted f and is the inverse of the wave's period (that is, $f=1/T$).

Phase: This is a given position in the cycle of the wave. It is most commonly used in discussing a "being out of phase" or a "phase shift", an offset between waves.

We can visualize several of these terms on the following diagram of a simple, periodic sine wave on a scale of distance displacement (x-axis) vs. (y-axis):

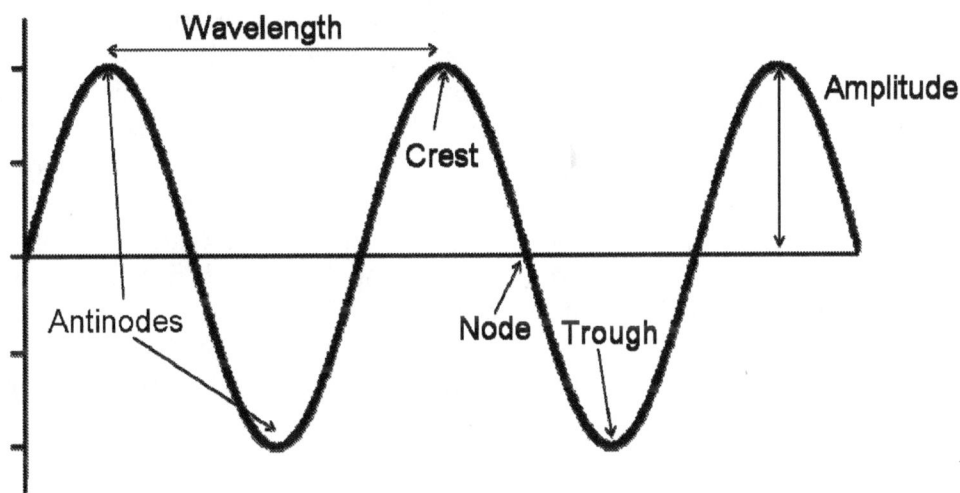

The phase velocity of a wave is related to its wavelength and frequency. Taking light waves, for instance, the speed of light c is equal to the distance traveled divided by time taken. Since the light wave travels the distance of one wavelength λ in the period of the wave T,

$$c = \frac{\lambda}{T}$$

The frequency of a wave, f, is the number of completed periods in one second. In general,

$$f = \frac{1}{T}$$

So the formula for the speed of light can be rewritten as

$$c = \lambda f$$

Thus the phase velocity of a wave is equal to the wavelength times the frequency.

Skill 18.3   Analyze factors determining the energy and power of a wave

Waves are an interesting energy transport phenomenon in that they carry energy but do not actually transport matter. Even when a wave moves through a medium, the individual particles of the medium are not transported from one place to another. An easily visualized example of this phenomenon involves a child's spring toy (such as a Slinky). By displacing the first coil, a person can introduce a pulse or wave into the toy. Thus, in doing this work, the person transfers energy to the medium (the Slinky) and this energy moves from one coil to the next and back again. If another person were to place their hand on the other end of the spring, they might feel the energy being transfer there. The energy carried by the wave is capable of doing work and, if it carried sufficient energy, it could displace the person's hand.

The amount of energy carried by a given wave is a function of its amplitude, as given by the following relationship:

$$E \propto A^2$$

That is, the energy is proportional to the square of the amplitude. For example, doubling the amplitude of a wave will quadruple the energy it carries. Returning to our analogy of the Slinky, we can see that this makes intuitive sense. To introduce energy to the Slinky, a person displaces the first coil. The greater energy they wish to add, the more they will displace this coil. This immediately results in greater amplitude. Note that this additional energy increases amplitude, but not frequency.

Finally, we must consider what role the medium plays in a wave's ability to transport energy. Considering our Slinky analogy once again, we can imagine two springs of different stiffness. In the stiffer spring, it will require more energy to cause the same displacement and create a wave with the same amplitude. We can extend this to understand the difference in waves traveling through any media with different elasticity. The more elastic the medium, the less resistance it will have to force. Thus a wave traveling through a more elastic medium will carry less energy than a wave of the same amplitude traveling through a less elastic medium.

Additionally, it is interesting to consider energy transformations in oscillating systems, which also provide us with a good example of the conservation of energy. Total energy must always be conserved, but in an oscillating system that energy is alternately in the form of kinetic and potential energy. Imagine, for instance, a mass connected to a spring. As the spring is compressed, elastic potential energy is stored within the spring. Specifically, when the spring is maximally compressed it also has maximum potential energy and minimum kinetic energy (zero kinetic energy). When the spring is released, this potential energy converts back into kinetic energy as the mass travels outward. This conversion back and forth happens continually as the spring oscillates.

For a mass on a spring with spring constant k, amplitude of oscillation A, angular frequency ω, time t, and phase Φ:

$$\text{Kinetic energy} = \tfrac{1}{2}\, k\, A^2 \sin^2(\omega_0 t + \Phi)$$

$$\text{Potential energy} = \tfrac{1}{2}\, k\, A^2 \cos^2(\omega_0 t + \Phi)$$

$$\text{Total energy} = \tfrac{1}{2}\, k\, A^2$$

We know that sine and cosine are 90° out of phase. Thus, this is a mathematical statement that when potential energy is maximal, kinetic energy is minimum. Further, kinetic energy plus potential energy is always equal to a constant, that is, total energy in the system is always conserved.

Skill 18.4 Apply the superposition principle to determine characteristics of a resultant wave

According to the principle of linear superposition, when two or more waves exist in the same place, the resultant wave is the sum of all the waves, i.e. the amplitude of the resulting wave at a point in space is the sum of the amplitudes of each of the component waves at that point. Interference is usually observed in coherent waves, well-correlated waves that have very similar frequencies or even come from the same source.

Superposition of waves may result in either constructive or destructive interference. Constructive interference occurs when the crests of the two waves meet at the same point in time. Conversely, destructive interference occurs when the crest of one wave and the trough of the other meet at the same point in time. It follows, then, that constructive interference increases amplitude and destructive interference decreased it. We can also consider interference in terms of wave phase; waves that are out of phase with one another will interfere destructively while waves that are in phase with one another will interfere constructively. In the case of two simple sine waves with identical amplitudes, for instance, amplitude will double if the waves are exactly in phase and drop to zero if the waves are exactly 180° out of phase.

Additionally, interference can create a standing wave, a wave in which certain points always have amplitude of zero. Thus, the wave remains in a constant position. Standing waves typically results when two waves of the same frequency traveling in opposite directions through a single medium are superposed.

View an animation of how interference can create a standing wave at the following URL:

http://www.glenbrook.k12.il.us/GBSSCI/PHYS/mmedia/waves/swf.html

All wavelengths in the EM spectrum can experience interference but it is easy to comprehend instances of interference in the spectrum of visible light. One classic example of this is Thomas Young's double-slit experiment. In this experiment a beam of light is shone through a paper with two slits and a striated wave pattern results on the screen. The light and dark bands correspond to the areas in which the light from the two slits has constructively (bright band) and destructively (dark band) interfered.

Similarly, we may be familiar with examples of interference in sound waves. When two sounds waves with slightly different frequencies interfere with each other, beat results. We hear a beat as a periodic variation in volume with a rate that depends on the difference between the two frequencies. You may have observed this phenomenon when listening to two instruments being tuned to match; beating will be heard as the two instruments approach the same note and disappear when they are perfectly in tune.

COMPETENCY 19.0 **Understand the principles of wave reflection, refraction, diffraction, interference, polarization, dispersion, and the Doppler effect**

Skill 19.1 **Analyze practical applications of wave reflection, refraction, diffraction, interference, polarization, dispersion, and the Doppler effect**

Wave refraction is a change in direction of a wave due to a change in its speed. This most commonly occurs when a wave passes from one material to another, such as a light ray passing from air into water or glass. However, light is only one example of refraction; any type of wave can undergo refraction. Another example would be physical waves passing from water into oil. At the boundary of the two media, the wave velocity is altered, the direction changes, and the wavelength increases or decreases. However, the frequency remains constant.

The index of refraction, n, is the amount by which light slows in a given material and is defined by the formula

$$n = \frac{c}{v}$$

where v represents the speed of light through the given material.

Problem: The speed of light in an unknown medium is measured to be $1.24 x10^8 m/s$. What is the index of refraction of the medium?

Solution:
$$n = \frac{c}{v}$$
$$n = \frac{3.00 x10^8}{1.24 x10^8} = 2.42$$

Referring to a standard table showing indices of refraction, we would see that this index corresponds to the index of refraction for diamond.

Reflection is the change in direction of a wave at an interface between two dissimilar media such that the wave returns into the medium from which it originated. The most common example of this is light waves reflecting from a mirror, but sound and water waves can also be reflected. The law of reflection states that the angle of incidence is equal to the angle of reflection.

Diffraction occurs when part of a wave front is obstructed. Diffraction and interference are essentially the same physical process. Diffraction refers to various phenomena associated with wave propagation such as the bending, spreading, and interference of waves emerging from an aperture. It occurs with any type of wave including sound waves, water waves, and electromagnetic waves such as light and radio waves.

Here, we take a close look at important phenomena like single-slit diffraction, double-slit diffraction, diffraction grating, other forms of diffraction and lastly interference.

1. **Single-slit diffraction:** The simplest example of diffraction is single-slit diffraction in which the slit is narrow and a pattern of semi-circular ripples is formed after the wave passes through the slit.
2. **Double-slit diffraction:** These patterns are formed by the interference of light diffracting through two narrow slits.
3. **Diffraction grating:** Diffraction grating is a reflecting or transparent element whose optical properties are periodically modulated. In simple terms, diffraction gratings are fine parallel and equally spaced grooves or rulings on a material surface. When light is incident on a diffraction grating, light is reflected or transmitted in discrete directions, called diffraction orders. Because of their light dispersive properties, gratings are commonly used in monochromators and spectrophotometers. Gratings are usually designated by their groove density, expressed in grooves/millimeter. A fundamental property of gratings is that the angle of deviation of all but one of the diffracted beams depends on the wavelength of the incident light.

4. **Other forms of diffraction:**

   i) Particle diffraction: It is the diffraction of particles such as electrons, which is used as a powerful argument for quantum theory. It is possible to observe the diffraction of particles such as neutrons or electrons and hence we are able to infer the existence of wave particle duality.

   ii) Bragg diffraction: This is diffraction from a multiple slits, and is similar to what occurs when waves are scattered from a periodic structure such as atoms in a crystal or rulings on a diffraction grating. Bragg diffraction is used in X-ray crystallography to deduce the structure of a crystal from the angles at which the X-rays are diffracted from it.

5. **Interference:** Interference is described as the superposition of two or more waves resulting in a new wave pattern. Interference is involved in Thomas Young's double slit experiment where two beams of light which are coherent with each other interfere to produce an interference pattern. Light from any source can be used to obtain interference patterns. For example, Newton's rings can be produced with sun light. However, in general, white light is less suited for producing clear interference patterns as it is a mix of a full spectrum of colors. Sodium light is close to monochromatic and is thus more suitable for producing interference patterns. The most suitable is laser light as it is almost perfectly monochromatic.

Dispersion is the separation of a wave into its constituent wavelengths due to interaction with a material occurring in a wavelength-dependent manner (as in thin-film interference for instance). Dispersive prisms separate white light into these constituent colors by relying on the differences in refractive index that result from the varying frequencies of the light. Prisms rely on the fact that light changes speed as it moves from one medium to another. This then causes the light to be bent and/or reflected. The degree to which bending/reflection occurs is a function of the light's angle of incidence and the refractive indices of the media.

Polarization is a property of transverse waves that describes the plane perpendicular to the direction of travel in which the oscillation occurs. In unpolarized light, the transverse oscillation occurs in all planes perpendicular to the direction of travel. Polarized light (created, for instance, by using polarizing filters that absorb light oscillating in other planes) oscillates in only a selected plane. An everyday example of polarization is found in polarized sunglasses which reduce glare.

The **Doppler effect** is the name given to the perceived change in frequency that occurs when the observer or source of a wave is moving. Specifically, the perceived frequency increases when a wave source and observer move toward each other and decreases when a source and observer move away from each other. Thus, the source and/or observer velocity must be factored in to the calculation of the perceived frequency. The mathematical statement of this effect is:

$$f' = f_0 \left(\frac{v \pm v_o}{v \pm v_s} \right)$$

where f' = observed frequency
f_0 = emitted frequency
v = the speed of the waves in the medium
v_s = the velocity of the source (positive in the direction away from the observer)
v_o = the velocity of the observer (positive in the direction towards the source)

Note that any motion that changes the perceived frequency of a wave will cause the Doppler effect to occur. Thus, the wave source, the observer position, or the medium through which the wave travels could possess a velocity that would alter the observed frequency of a wave.

So, let's consider two examples involving sirens and analyze what happens when either the source or the observer moves. First, imagine a person standing on the side of a road and a police car driving by with its siren blaring. As the car approaches, the velocity of the car will mean sound waves will "hit" the observer as the car comes closer and so the pitch of the sound will be high. As it passes, the pitch will slide down and continue to lower as the car moves away from the observer. This is because the sound waves will "spread out" as the source recedes. Now consider a stationary siren on the top of fire station and a person driving by that station. The same Doppler effect will be observed: the person would hear a high frequency sound as he approached the siren and this frequency would lower as he passed and continued to drive away from the fire station.

Problem: A sound source moves towards an observer at 8m/s while the observer moves towards the source at 9m/s. If the observer detects a sound of frequency 83kHz, what is the frequency of the sound emitted? The speed of sound in the medium is 343 m/s.

Solution: Using the Doppler equation, the emitted frequency f is given by

$$83 = f \frac{343 + 9}{343 - 8}; \Rightarrow f = 79kHz$$

Skill 19.2   Apply Snell's law to determine index of refraction, angle of refraction, or critical angle as a wave passes from one medium to another

Snell's Law describes how light bends, or refracts, when traveling from one medium to the next. It is expressed as

$$n_1 \sin \theta_1 = n_2 \sin \theta_2$$

where n_i represents the index of refraction in medium i, and θ_i represents the angle the light makes with the normal in medium i.

Problem:

The index of refraction for light traveling from air into an optical fiber is 1.44.

(a) In which direction does the light bend?

(b) What is the angle of refraction inside the fiber, if the angle of incidence on the end of the fiber is 22°?

Solution: (a) The light will bend toward the normal since it is traveling from a rarer region (lower n) to a denser region (higher n).

(b) Let air be medium 1 and the optical fiber be medium 2:

$$n_1 \sin \theta_1 = n_2 \sin \theta_2$$
$$(1.00) \sin 22° = (1.44) \sin \theta_2$$
$$\sin \theta_2 = \frac{1.00}{1.44} \sin 22° = (.6944)(.3746) = 0.260$$
$$\theta_2 = \sin^{-1}(0.260) = 15°$$

The angle of refraction inside the fiber is $15°$.

Reflection may occur whenever a wave travels from a medium of a given refractive index to another medium with a different index. A certain fraction of the light is reflected from the interface and the remainder is refracted. However, when the wave is moving from a dense medium into one less dense, that is the refractive index of the first is greater than the second, a **critical angle** exists which will create a phenomenon known as **total internal reflection.** In this situation all of the wave is reflected. The critical angle of incidence θ_c is the one for which the angle of reflection is 90 degrees. Thus, according to Snell's law

$$n_1 \sin \theta_c = n_2 \sin 90^0 \Rightarrow \theta_c = \sin^{-1} \frac{n_2}{n_1}$$

Snell's law may also be used to understand the phenomenon of dispersion since it relates the angle of refraction at the boundary between two media to the relative refractive indices. Since different frequency components of visible light have different indices of refraction in any medium other than vacuum, each component of a beam of white light is refracted at a different angle when it crosses a surface resulting in a separation of the colors.

Skill 19.3   Analyze problems involving diffraction and interference in single and multiple slits

All wavelengths in the EM spectrum can experience interference but it is easy to comprehend instances of interference in the spectrum of visible light. Please see **Skill 19.1** for an introduction to interference and diffraction. One classic example of this phenomenon is **Thomas Young's double-slit experiment**. In this experiment a beam of light is shone through a paper with two slits and a striated wave pattern results on the screen. The light and dark bands correspond to the areas in which the light from the two slits has constructively (bright band) and destructively (dark band) interfered. Light from any source can be used to obtain interference patterns. For example, Newton's rings can be produced with sun light. However, in general, white light is less suited for producing clear interference patterns as it is a mix of a full spectrum of colors. Sodium light is close to monochromatic and is thus more suitable for producing interference patterns. The most suitable is laser light as it is almost perfectly monochromatic.

Problem: The interference maxima (location of bright spots created by constructive interference) for double-slit interference are given by

$$\frac{n\lambda}{d} = \frac{x}{D} = \sin\theta \quad n=1,2,3...$$

where λ is the wavelength of the light, d is the distance between the two slits, D is the distance between the slits and the screen on which the pattern is observed and x is the location of the nth maximum. If the two slits are 0.1mm apart, the screen is 5m away from the slits, and the first maximum beyond the center one is 2.0 cm from the center of the screen, what is the wavelength of the light?

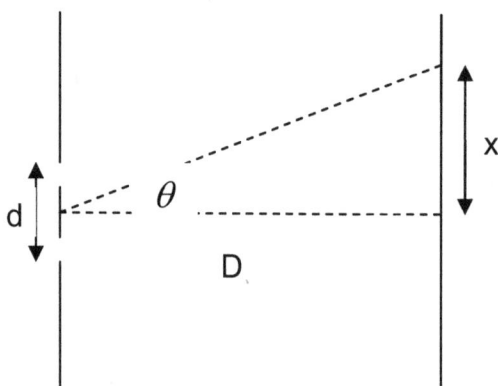

Solution: λ = xd/(Dn) = 0.02 x 0.0001/ (5 x1) = 400 nanometers

For **single-slit diffraction**, a central bright fringe of light is observed since waves from all points in the slit travel approximately the same distance to the center of the pattern and are in phase there. An interference pattern of alternate bright and dark fringes is observed around the central spot.

The location of the dark fringes is given by

$$a \sin\theta = m\lambda \quad \text{for } m = 1,2,3,\dots$$

where a is the width of the slit and λ is the wavelength of the light.

Problem:
The first minimum for light of wavelength 650 nm diffracted from a single slit is found at θ = 15 degrees. What is the width of the slit?

Solution:

Since m = 1 at the first minimum, $a = \dfrac{\lambda}{\sin\theta} = \dfrac{650}{\sin 15^0} = 2.5\mu m$.

COMPETENCY 20.0 **Understand the characteristics of sound waves and the means by which sound waves are produced and transmitted**

Skill 20.1 **Analyze the energy, power, and intensity of sound waves**

Sound waves are mechanical waves that can travel through various kinds of media, solid, liquid and gas. They are longitudinal waves transmitted as variations in the pressure of the surrounding medium. These variations in pressure form waves that are termed sound waves or acoustic waves. When the particles of the medium are drawn close together it is called compression. When the particles of the medium are spread apart it is known as rarefaction.

Sound waves have different characteristics in different materials. Boundaries between different media can result in partial or total reflection of sound waves. Thus, the phenomenon of echo takes place when a sound wave strikes a material of differing characteristics than the surrounding medium. The reflection of a voice off the walls of a room is a particular example.

The frequency of the harmonic acoustic wave, ω, determines the pitch of the sound in the same manner that the frequency determines the color of an electromagnetic harmonic wave. A high pitch sound corresponds to a high frequency sound wave and a low pitch sound corresponds to a low frequency sound wave. Similarly, the amplitude of the pressure determines the loudness or just as the amplitude of the electric (or magnetic) field E determines the brightness or intensity of a color. The intensity of a sound wave is proportional to the square of its amplitude.

The decibel scale is used to measure sound intensity. It originated in a unit known as the bel which is defined as the reduction in audio level over 1 mile of a telephone cable. Since the bel describes such a large variation in sound, it became more common to use the decibel, which is equal to 0.1 bel. A decibel value is related to the intensity of a sound by the following equation:

$$X_{dB} = 10\log_{10}\left(\frac{X}{X_0}\right)$$

Where X_{dB} is the value of the sound in decibels
   X is the intensity of the sound
   X_0 is a reference value with the same units as X. X_0 is commonly taken to
      be the threshold of hearing at $10^{-12} W/m^2$.

It is important to note the logarithmic nature of the decibel scale and what this means for the relative intensity of sounds. The perception of the intensity of sound increases logarithmically, not linearly. Thus, an increase of 10 dB corresponds to an increase by one order of magnitude.

For example, a sound that is 20 dB is not twice as loud as sound that is 10 dB; rather, it is 10 times as loud. A sound that is 30 dB will the 100 times as loud as the 10 dB sound.

Finally, let's equate the decibel scale with some familiar noises. Below are the decibel values of some common sounds.

Whispering voice: 20 dB
Quiet office: 60 dB
Traffic: 70 dB
Cheering football stadium: 110 dB
Jet engine (100 feet away): 150 dB
Space shuttle liftoff (100 feet away): 190 dB

Skill 20.2   Analyze factors that affect the speed of sound in different media

Mechanical waves rely on the local oscillation of each atom in a medium, but the material itself does not move; only the energy is transferred from atom to atom. Therefore the material through which the mechanical wave is traveling greatly affects the wave's propagation and speed. In particular, a material's elastic constant and density affect the speed at which a wave travels through it. Both of these properties of a medium can predict the extent to which the atoms will vibrate and pass along the energy of the wave. The general relationship between these properties and the speed of a wave in a solid is given by the following equation:

$$V = \sqrt{\frac{C_{ij}}{\rho}}$$

where V is the speed of sound, C_{ij} is the elastic constant or bulk modulus, and ρ is the material density. It is worth noting that the elastic constant differs depending on direction in anisotropic materials and the ij subscripts indicate that this directionality must be taken into account.

The speed of sound also varies with temperature since material characteristics are often dependent on temperature. At $0^0 C$, the speed of sound is 331m/s in air, 1402m/s in water and 6420m/s in Aluminum. Even though water is much denser than air, it is a lot more incompressible than air and its bulk modulus is larger than that of air by a larger factor. Therefore sound travels much faster in water than in air.

In the case of a stretched string, the velocity of a transverse wave passing through it is given by

$$v = \sqrt{\frac{\tau}{\mu}}$$

where τ is the tension in the string and μ is its linear density.

Skill 20.3 Solve problems involving resonance, harmonics, and overtones in vibrating strings and air columns

A standing wave typically results from the interference between two waves of the same frequency traveling in opposite directions. The result is a stationary vibration pattern. One of the key characteristics of standing waves is that there are points in the medium where no movement occurs. The points are called nodes and the points where motion is maximal are called antinodes. Thus the amplitude of a standing wave varies with location. This property allows for the analysis of various typical standing waves.

Strings
Imagine a string of length L tied tightly at its two ends. We can generate a standing wave by plucking the string. The waves traveling along the string are reflected at the fixed end points and interfere with each other to produce standing waves. There will always be two nodes at the ends where the string is tied. Depending on the frequency of the wave that is generated, there may also be other nodes along the length of the string. In the diagrams below, examples are given of strings with 0, 1, or 2 additional nodes.

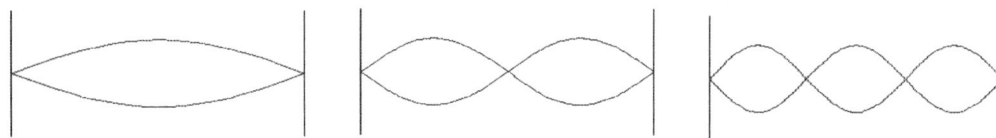

Percussion
On a percussion instrument such as a drum, standing waves occur on the two-dimensional surface in many different configurations. The nodes are circles and straight lines rather than points as on a string. Some possible node patterns are shown below.

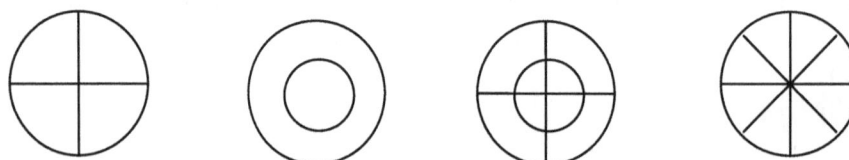

Winds
Just as on a string, standing waves can propagate in gaseous or liquid medium inside a tube. In these cases we will also observe harmonic vibrations, but their nature will depend on whether the ends of the tube are closed or open. Specifically, an antinode will be observed at an open end and a node will appear at a closed end. As in strings, there may be additional nodes in between.

Due to the presence of nodes in a standing wave pattern, a stretched string or a tube with ends closed vibrates at certain characteristic frequencies that correspond to different node patterns. These frequencies are known as **harmonics** or characteristic **resonance frequencies** of the instrument. The instrument will absorb much greater energy when excited at one of the resonance frequencies. The lowest of these frequencies is known as the **fundamental** while the higher ones are known as **overtones**. Harmonic frequencies for vibrating strings and air tubes are calculated below.

Vibrating string
Since we know the length of the string (L) in each case, we can calculate the wavelength (λ) and frequency (f) for any harmonic using the following formula, where n=the harmonic order (n=1,2,3...) and v is the phase velocity of the wave.

$$\lambda_n = \frac{2L}{n} \qquad\qquad f_n = \frac{v}{\lambda_n} = n\frac{v}{2L}$$

Waves in a tube
For a tube with two closed ends or two open ends (note that this is the same as for the string described above):

$$f_n = \frac{nv}{2L}$$

where n = 1,2,3,4...

When only one end of a tube is closed, that end become a node and wave exhibits odd harmonics. For a tube with one close end and one open end:

$$f_n = \frac{nv}{4L}$$

where n = 1,3,5,7...

Wavelengths can be found by applying the formula f=v/λ.

Problem:

A string on a violin is 75 cm long. The speed of waves produced when the string is played is known to be 345 m/s. What is the fundamental frequency of the string?

Solution:

Remember that fundamental frequency is simply the first harmonic of the string. The wavelength of the first harmonic must be twice the length of the string:

$$\lambda = 2 \times l = 2 \times 0.75m = 1.5m$$

Then find the fundamental frequency:

$$f = \frac{v}{\lambda} = \frac{345\frac{m}{s}}{1.5m} = 230s^{-1} = 230Hz$$

COMPETENCY 21.0   Understand the production and characteristics of electromagnetic waves

Skill 21.1   Analyze the properties and components of the electromagnetic spectrum

The electromagnetic spectrum is measured using frequency (f) in hertz or wavelength (λ) in meters. The frequency times the wavelength of every electromagnetic wave equals the speed of light (3.0×10^8 meters/second).

Roughly, the range of wavelengths of the electromagnetic spectrum is:

	f	**λ**
Radio waves	$10^{5} - 10^{-1}$ hertz	10^{3} -10^{9} meters
Microwaves	3×10^{9} - 3×10^{11} hertz	10^{-3} -10^{-1} meters
Infrared radiation	3×10^{11} - 4×10^{14} hertz	7×10^{-7} -10^{-3} meters
Visible light	4×10^{14} -7.5×10^{14} hertz	4×10^{-7} -7×10^{-7} meters
Ultraviolet radiation	7.5×10^{14} -3×10^{16} hertz	10^{-8} -4×10^{-7} meters
X-Rays	3×10^{16} - 3×10^{19} hertz	10^{-11} -10^{-8} meters
Gamma Rays	$>3 \times 10^{19}$ hertz	$<10^{-11}$ meters

Radio waves are used for transmitting data. Common examples are television, cell phones, and wireless computer networks. Microwaves are used to heat food and deliver Wi-Fi service. Infrared waves are utilized in night vision goggles. Visible light we are all familiar with as the human eye is most sensitive to this wavelength range. UV light causes sunburns and would be even more harmful if most of it were not captured in the Earth's ozone layer. X-rays aid us in the medical field and gamma rays are most useful in the field of astronomy.

Skill 21.2   Analyze the energy, frequency, and amplitude of electromagnetic waves in terms of the vibrations of the sources that produce them

Vibrations, as opposed to unidirectional motion, require an unbalancing force followed by a restoring force. A common macroscopic example is a spring harmonic oscillator with a mass suspended from a stationary surface by way of a spring. If the mass is pulled down further (an unbalancing force), the tension of the spring increases as the mass is displaced until an equilibrium is reached. Once released, the mass will return to its original location; at this point, however, the energy used to displace the mass has been converted from potential energy to kinetic energy. Although the forces are balanced at the point of zero initial displacement, the kinetic energy of the mass causes it to pass the initial location, thus converting kinetic energy to potential energy once more as a restoring force increases in an effort to return the mass to its initial location. This process repeats, resulting in oscillations.

A similar process can take place at the microscopic level with atoms or molecules in a material. As the frequency and amplitude of the constituent elements of the material increase, the result is a corresponding increase in thermal energy. This increase in energy can be registered as an increase in temperature.

When charged particles or molecules acquire vibrational energy, the energy can be radiated as electromagnetic waves (or photons). Electrons are a simple example. An ensemble of electrons in a metal rod, for instance, can vibrate as the result of an applied alternating electric field. This causes an alternating current, which, according to Maxwell's equations, leads to a propagating electromagnetic field. The frequency of oscillation determines the frequency of the wave. If the charged particle is in a bulk material, the energy from oscillation can either be emitted as electromagnetic radiation, as mentioned, or as a phonon, a vibrational mode that is treated like a particle. This is a more refined way of addressing thermal energy, but is similar in terms of the basic concepts.

From a quantum mechanical perspective, oscillations (such as those of electrons) can be dealt with by solving the Schrödinger wave equation using the quantum mechanical Hamiltonian. This Hamiltonian is similar in form to the expression for the mechanical energy of a classical oscillator and includes a term for kinetic energy and a term for potential energy. Instead of producing a continuous spectrum of possible energies, however, the quantum mechanical approach yields a discrete set of energy levels for the oscillator. Changes between energy levels require the emission (or absorption) of a specific discrete amount of energy. Such changes can occur through the emission or absorption of a photon or phonon.

Skill 21.3   Analyze practical applications of electromagnetic waves

The characteristics of electromagnetic radiation vary depending on the frequency of the radiation and the properties of the material through which it is propagating. Virtually innumerable applications have been devised for various frequency bands, leading to the development of a number of interesting and practical devices and inventions.

A common application of the microwave portion of the electromagnetic spectrum is the microwave oven. Since water molecules have a local charge imbalance, they each manifest a dipole moment. Microwave radiation operates at a wavelength such that incidence on water molecules imparts oscillating motion because of the dipole moments. This motion causes the temperature of the water to rise, thus cooking the food.

Another useful application of the electromagnetic spectrum is remote communications. A number of different frequency bands can be exploited, each with its own characteristics. Since oscillation of charges produces electromagnetic waves that propagate away from the source, antennas that are supplied with an electrical signal can be used to send or receive information through modulation of the waves. Two common modulation schemes are amplitude modulation (AM), for which the amplitude of a sine wave at a given frequency is varied, and frequency modulation (FM), for which the frequency of a sine wave (with a specific central frequency) is modulated. In both cases, these modulation schemes use a band of frequencies (bandwidth) in accordance with Fourier theory. Thus, a broadcast radio station is actually allocated a range of frequencies around a single central frequency (50 kHz around 100.5 MHz, for example). A virtually unlimited range of frequencies can be used for communication although some are more suited to certain situations than others. For example, line-of-sight communication is amenable to the use of a visible (or non-visible) laser as well as microwaves which are commonly used for cellular phones.

In cases where visible light is dim or absent, infrared detectors or video equipment can be used to produce images of scenes that appear dark to the unaided eye. Depending on their temperature, objects radiate infrared waves which can be used to produce false-color images of an otherwise dark environment.

Another method of imaging is the use of X-rays. These high-energy (i.e., high frequency) waves are able to penetrate certain materials, such as flesh, but are unable to penetrate other materials, such as bone or metal. Thus, X-rays may be used to visualize the skeletal structure of a human or animal without an invasive procedure or they may be used to detect hidden metallic items, such as various forms of contraband.

An advanced form of imaging is holography, which allows recording of a three-dimensional image, rather than just a two-dimensional image. Holography uses the coherent light produced by lasers to imprint an image on a photographic plate; in addition to amplitude information, which is all that is recorded in standard photography, holography also includes phase information, thus allowing capture of a three-dimensional image.

The broad range of frequencies radiated by the Sun can be collected in a number of ways to beneficially use or convert the incident energy. So-called solar panels can be used to convert electromagnetic radiation into electricity for powering devices or for storage in batteries. Also, the heat produced as a result of absorption of solar radiation can be used, for example, to heat a house. An appropriate system to collect the radiation, convert it to heat and distribute it throughout a house or other structure is one particular use of this phenomenon.

COMPETENCY 22.0   Understand and apply the principles of lenses and mirrors

Skill 22.1   Compare types and characteristics of lenses and mirrors

Lenses are devices that causes electromagnetic radiation to converge or diverge. The most familiar lenses are made of glass or plastic and designed to concentrate or disperse visible light. Two of the most important parameters for a lens are its thickness and its focal length. Focal length is a measure of how strongly light is concentrated or dispersed by a lens. For a convex or converging lens, the focal length is the distance at which a beam of light will be focused to a single spot. Conversely, for a concave or diverging lens, the focal length is the distance to the point from which a beam appears to be diverging.

The images produced by lenses can be either virtual or real. A virtual image is one that is created by rays of light that appear to diverge from a certain point. Virtual images cannot be seen on a screen because the light rays do not actually meet at the point where the image is located. If an image and object appear on the same side of a converging lens, that image is defined as virtual. For virtual images, the image location will be negative and the magnification positive. Real images, on the other hand, are formed by light rays actually passing through the image. Thus, real images are visible on a screen. Real images created by a converging lens are inverted and have a positive image location and negative magnification.

Plane mirrors form virtual images. In other words, the image is formed behind the mirror where light does not actually reach. The image size is equal to the object size and object distance is equal to the image distance; i.e. the image is the same distance behind the mirror as the object is in front of the mirror. Another characteristic of plane mirrors is left-right reversal. For example, suppose you are standing in front of a mirror with your right hand raised. The image in the mirror will be raising its left hand.

Problem: If a cat creeps toward a mirror at a rate of 0.20 m/s, at what speed will the cat and the cat's image approach each other?

Solution: In one second, the cat will be 0.20 meters closer to the mirror. At the same time, the cat's image will be 0.20 meters closer to the cat. Therefore, the cat and its image are approaching each other at the speed of 0.40 m/s.

Problem: If an object that is two feet tall is placed in front of a plane mirror, how tall will the image of the object be?

Solution: The image of the object will have the same dimensions as the actual object, in this case, a height of two feet. This is because the magnification of an image in a plane mirror is 1.

Curved mirrors

Curved mirrors are usually sections of spheres. In a concave mirror the inside of the spherical surface is silvered while in a convex mirror it is the outside of the spherical surface that is silvered.

Terminology associated with spherical mirrors:

Principal axis: The line joining the center of the sphere (of which we imagine the mirror is a section) to the center of the reflecting surface.

Center of curvature: The center of the sphere of which the mirror is a section.

Vertex: The point on the mirror where the principal axis meets the mirror or the geometric center of the mirror.

Focal point: The point at which light rays traveling parallel to the principal axis will meet after reflection in a concave mirror. For a convex mirror, it is the point from which light rays traveling parallel to the principal axis will appear to diverge from after reflection. The focal point is midway between the center of curvature and the vertex.

Focal length: The distance between the focal point and the vertex.

Radius of curvature: The distance between the center of curvature and the vertex, i.e. the radius of the sphere of which the mirror is a section. The radius of curvature is twice the focal length.

Image characteristics for **concave mirrors**:

1. If the object is located beyond the center of curvature, the image will be real, inverted, smaller and located between the focal point and center of curvature.
2. If the object is located at the center of curvature, the image will be real, inverted, of the same height and also located at the center of curvature.
3. If the object is located between the center of curvature and focal point, the image will be real, inverted, larger and located beyond the center of curvature.
4. If the object is located at the focal point no image is formed.
5. If the object is located between the focal point and vertex, the image will be virtual, upright, larger and located on the opposite side of the mirror.
6. If the object is located at infinity (very far away), the image is real, inverted, smaller and located at the focal point.

For convex mirrors, the image is always virtual, upright, reduced in size and formed on the opposite side of the mirror.

Skill 22.2   Use a ray diagram to locate the focal point or point of image formation of a lens or mirror

Ray diagrams are a convenient way to visualize the propagation of waves and to perform reasonably accurate calculations of the effect of mirrors and lenses on light.

For mirrors, the focal point is either real (for concave mirrors) or virtual (for convex mirrors). In either case, the focal point is found by looking at the behavior of two parallel rays incident upon the mirror.

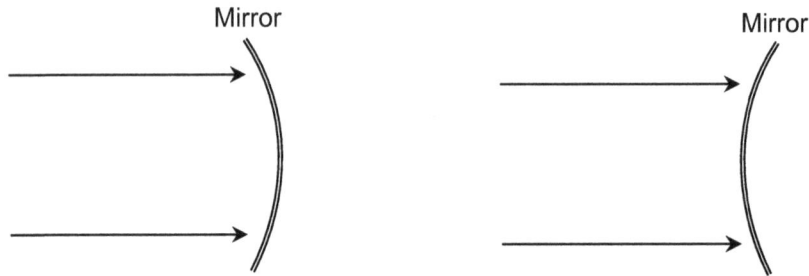

For each incident ray, the angle of reflection θ_r is equal to the angle of incidence θ_r, where the angle is measured from the normal to the mirror surface at the point of incidence.

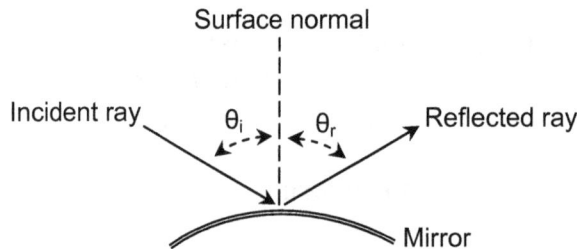

When this law is applied to the rays for either mirror, the focal points (f) are revealed as the (real or virtual) intersection of the rays. The virtual focus point is the intersection for the reflected rays when they are extended beyond the surface of the mirror.

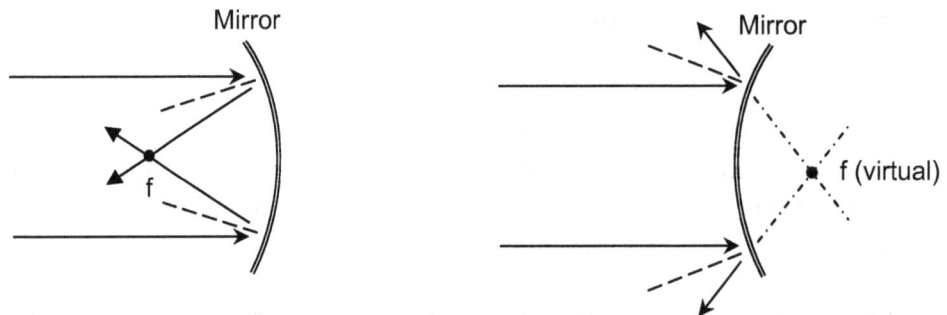

The focal point of a lens is found in a similar manner, with the exception that instead of using the law of reflection, refraction by way of Snell's law must be applied. Since real lenses have a finite thickness, Snell's law must be used for the ray both as it enters the lens material and as it exits the lens material.

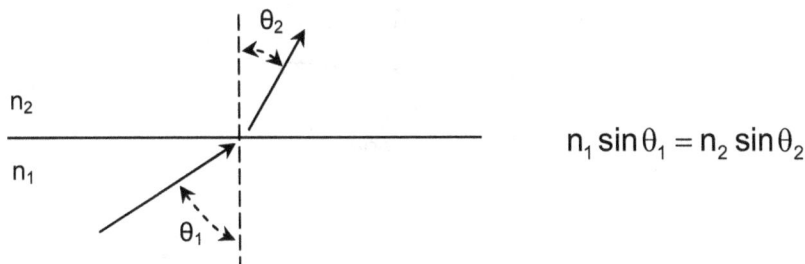

$$n_1 \sin \theta_1 = n_2 \sin \theta_2$$

As with reflection, refraction requires consideration of the normal to the surface. Also, the refractive indices must be used. It is assumed here that the refractive index of the outer medium is n_1 and the refractive index of the lens is n_2, and that $n_1 < n_2$. The intersection of parallel incident rays determines the focal point, which may again be either real or virtual. Real focal points occur on the opposite side of the lens from the source of illumination, and virtual focal points occur on the same side as the source of illumination.

Only one lens case is shown here, but the principles apply equally to all variations of lenses.

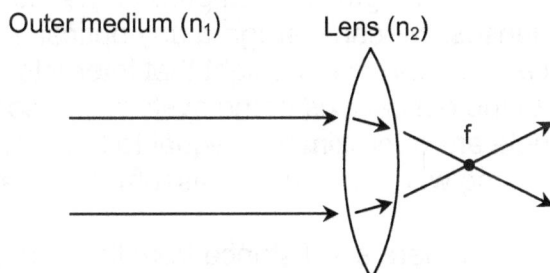

Care must be taken in properly identifying the normal and in applying Snell's law to both incidences of refraction for each ray.

Determining the point of image formation for mirrors and lenses is accomplished in a similar manner to that of the focal points. As with the focal points, image points may be either real or virtual, depending on the characteristics of the mirror or lens. To find the image point, two rays of differing angles must be traced as they interact with the lens or mirror. The initial directions of the rays can be chosen arbitrarily, but it is ideal to choose the directions such that the difficulty with determining the direction of the reflected or refracted ray is minimized. The intersection of these two rays is the image point. Only two examples are shown here, but the principles behind these examples may be applied to any variation of the situations, as well as to any combination of lenses and mirrors.

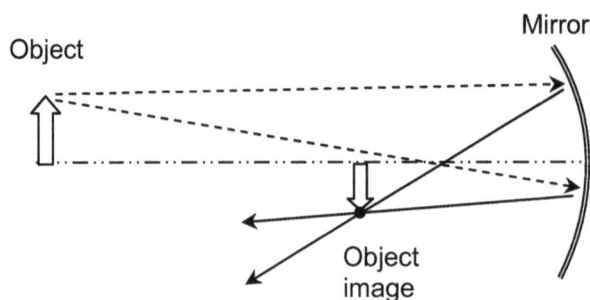

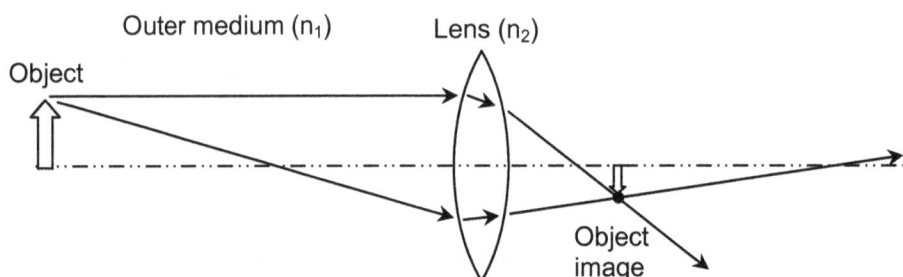

Skill 22.3   Apply the thin lens equation to solve problems involving lenses and mirrors

A thin lens in one in which focal length is much greater than lens thickness. For problems involving thin lenses, we can disregard any optical effects of the lens itself. Additionally, we can assume that the light that interacts with the lens makes a small angle with the optical axis of the system and so the sine and tangent values of the angle are approximately equal to the angle itself. This paraxial approximation, along with the thin lens assumptions, allows us to state:

$$\frac{1}{s} + \frac{1}{s'} = \frac{1}{f}$$

Where s = distance from the lens to the object
s' = distance from the lens to the image
f = focal length of the lens

Most lenses also cause some magnification of the object. Magnification is defined as:

$$m = \frac{y'}{y} = -\frac{s'}{s}$$

Where m = magnification
y' = image height
y = object height

Sign conventions will make it easier to understand thin lens problems:

Focal length: positive for a converging lens; negative for a diverging lens

Object location: positive when in front of the lens; negative when behind the lens

Image location: positive when behind the lens; negative when in front of the lens

Image height: positive when upright; negative when upside-down.

Magnification: positive for an erect, virtual image; negative for an inverted, real image

Problem:
A converging lens has a focal length of 10.00 cm and forms a 2.0 cm tall image of a 4.00 mm tall real object to the left of the lens. If the image is erect, is the image real or virtual? What are the locations of the object and the image?

Solution:
We begin by determining magnification:

$$m = \frac{y'}{y} = \frac{0.02m}{0.004m} = 5$$

Since the magnification is positive and the image is erect, we know the image must be virtual. To find the locations of the object and image, we first relate them by using the magnification:

$$m = -\frac{s'}{s}$$
$$s' = -ms$$

Then we substitute into the thin lens equation, creating one variable in one unknown:

$$\frac{1}{s} + \frac{1}{s'} = \frac{1}{f}$$

$$\frac{1}{s} - \frac{1}{5s} = \frac{1}{10cm}$$

$$\frac{5-1}{5s} = \frac{1}{10cm}$$

$$s = \frac{40cm}{5} = 8cm \rightarrow \rightarrow s' = -5 \times 8cm = -40cm$$

Thus the object is located 8 cm to the left of the lens and the image is 40 cm to the left of the lens.

The relationship between the object distance from vertex s, the image distance from vertex s', and the focal length f is given by the equation

$$\frac{1}{s} + \frac{1}{s'} = \frac{1}{f}$$

Magnification is defined as:

$$m = \frac{y'}{y} = -\frac{s'}{s}$$

where m=magnification, y'=image height, y=object height

Problem: A concave mirror collects light from a star. If the light rays converge at 50 cm, what is the radius of curvature of the mirror?

Solution: The point at which the rays converge is known as the focal point. The focal length, in this case, 50 cm, is the distance from the focal point to the mirror. The radius of curvature is the distance from the vertex to the center of curvature. The vertex is the point on the mirror where the principal axis meets the mirror. The center of curvature represents the point in the center of the sphere from which the mirror was sliced. Since the focal point is the midpoint of the line from the vertex to the center of curvature, or focal length, the focal length would be one-half the radius of curvature. Since the focal length in this case is 50 cm, the radius of curvature would be 100 cm.

Problem: An image of an object in a mirror is upright and reduced in size. In what type of mirror is this image being viewed, plane, concave, or convex?

Solution: The image in a plane mirror would be the same size as the object. The image in a concave mirror would be magnified if upright. Only a convex mirror would produce a reduced upright image of an object.

Skill 22.4   Analyze applications of lenses and mirrors

Eye

The eye is a very complex sensory organ. Although there are many critical anatomical features of the eye, the lens and retina are the most important for focusing and sensing light. Light passes through the cornea and through the lens. The lens is attached to numerous muscles that contract and relax to move the lens in order to focus the light onto the retina. The pupil also contracts and relaxes to allow more or less light in the eye as required.

The retina contains rod cells which are responsible for vision in low light and cone cells which sense color and detail. Different types of cone cells are capable of sensing different wavelengths of light. A chemical called rhodopsin is present in the retina that converts light signals into electrical impulses that are sent to the brain to interpret as vision. The retina is lined with a black pigment called melanin that reduces reflection.

The eye relies on refraction to focus light onto the retina. Refraction occurs at four curved interfaces; between the air and the front of the cornea, the back of the cornea and the aqueous humor, the aqueous humor at the front of the lens, and the back of the lens and the vitreous humor. When each of these interfaces are working properly the light arrives at the retina in perfect focus for transmission to the brain as an image.

Eye Glasses

When all the parts of the eye are not working together correctly, corrective lenses or eyeglasses may be needed to assist the eye in focusing the light onto the retina. The surfaces of the lens or cornea may not be smooth causing the light to refract in the wrong direction. This is called astigmatism. Another common problem is that the lens is not able to change its curvature appropriately to match the image. The cornea can also be misshaped resulting in blurred vision. Corrective lenses consist of curves pieces of glass which bend the light in order to change the focal point of the light. A nearsighted eye forms images in front of the retina. To correct this, a minus lens consisting of two concave prisms is used to bend light out and move the image back to the retina. A farsighted eye creates images behind the retina. This is corrected using plus lenses that bend light in and bring the image forward onto the retina. The worse the vision, the farther out of focus the image is on the retina. Therefore the stronger the lens the further the focal point is moved to compensate.

Spectroscope

Spectrometers known as spectroscopes are used to identify materials. Spectroscopes are used often in astronomy and some branches of chemistry. Early spectroscopes were simply a prism with graduations marking wavelengths of light. Modern spectroscopes typically use a diffraction grating, a movable slit, and some kind of photo detector, all automated and controlled by a computer. When materials are heated they emit light that is characteristic of its atomic composition. The emission of certain frequencies of light produce a pattern of lines that are comparable to a fingerprint. The yellow light emission of heated sodium is a typical example.

A spectroscope is able to detect, measure and record the frequencies of the emitted light. This is done by passing the light though a slit to a collimating lens which transforms the light into parallel rays. The light is then passed through a prism that refracts the beam into a spectrum of different wavelengths. The image is then viewed alongside a scale to determine the characteristic wavelengths.

Camera

A camera is another device that utilizes the lens' ability to refract light to capture and process an image. As with the eye, light enters the lens of a camera and focuses the light on the other side. Instead of focusing on the retina, the image is focused on the film to create a film negative. This film negative is later processed with chemicals to create a photograph. A camera uses a converging or convex lens. This lens captures and directs light to a single point to create a real image on the surface of the film. To focus a camera on an image, the distance of the lens from the film is adjusted in order to ensure that the real image converges on the surface of the film and not in front of or behind it.

Different lenses are available which capture and bend the light to different degrees. A lens with more pronounced curvature will be able to bend the light more acutely causing the image to converge more closely to the lens. Conversely a flatter lens will have a longer focal distance. The further the lens is located from the film (flatter lens), the larger the image becomes. Thus zoom lenses on cameras are flat while wide angle lenses are more rounded. The focal length number on a certain lens conveys the magnification ability of the lens.

The film functions like the retina of the eye in that it is light sensitive and can capture light images when exposed. However, this exposure must be brief to capture the contrasting amounts of light and a clear image. The rest of the camera functions to precisely control how much light contacts the film. The aperture is the lens opening which can open and close to let in more or less light. The temporal length of light exposure is controlled by the shutter which can be set at different speeds depending on the amount of action and level of light available. The film speed refers to the size of the light sensitive grains on the surface of the film. The larger grains absorb more light photons than the smaller grains, so film speed should be selected according to lighting conditions.

Telescope

A telescope is a device that has the ability to make distant objects appear to be much closer. Most telescopes are one of two varieties, a refractor which uses lenses or a reflector which uses mirrors. Each accomplishes the same purpose but in totally different ways. The basic idea of a telescope is to collect as much light as possible, focus it, and then magnify it. The objective lens or primary mirror of a telescope brings the light from an object into focus. An eyepiece lens takes the focused light and "spreads it out" or magnifies it using the same principle as a magnifying glass using two curved surfaces to refract the light.

Microscope

Microscopes are used to view objects that are too small to be seen with the naked eye. A microscope usually has an objective lens that collects light from the sample and an eyepiece which brings the image into focus for the observer. It also has a light source to illuminate the sample. Typical optical microscopes achieve magnification of up to 1500 times.

SUBAREA V. QUANTUM THEORY AND THE ATOM

COMPETENCY 23.0 Understand the photoelectric effect, quantum theory, and the dual nature of light and matter

Skill 23.1 Apply the laws of photoelectric emission to explain photoelectric phenomena

Einstein used the photon hypothesis to explain the photoelectric effect which is the emission of electrons from a metal surface when light is incident on it. When this metal surface is a cathode with the anode held at a higher potential V, the emitted electrons create a current flow in the external circuit. It is observed that current flows only for light of higher frequencies, i.e. electrons are released from the metal only by high frequency light. Also there is a threshold negative potential, the stopping potential V_0 below which no current will flow in the circuit.

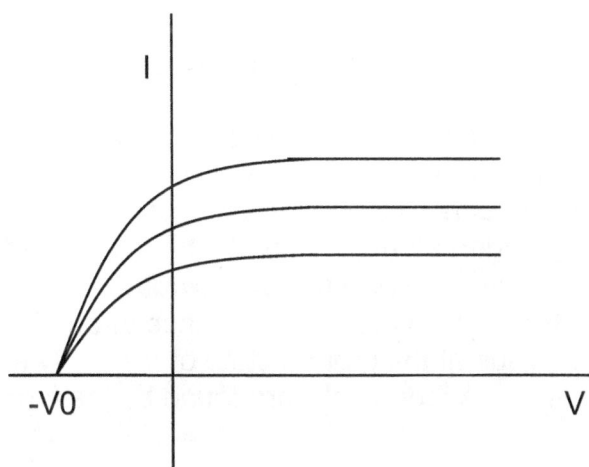

The figure displayed above shows current flow vs. potential for three different intensities of light. It shows that the maximum current flow increases with increasing light intensity but the stopping potential remains the same.

All these observations are counter-intuitive if one considers light to be a wave but may be understood in terms of light particles or photons. According to this interpretation, each photon transfers its energy to a single electron in the metal. Since the energy of a photon depends on its frequency, only a photon of higher frequency can transfer enough energy to an electron to enable it to pass the stopping potential threshold.

When V is negative, only electrons with a kinetic energy greater than $|eV|$ can reach the anode. The maximum kinetic energy of the emitted electrons is given by eV_0. This is expressed by Einstein's photoelectric equation as

$$(\tfrac{1}{2}mv^2)_{max} = eV_0 = hf - \varphi$$

where the **work function** φ is the energy needed to release an electron from the metal and is characteristic of the metal.

Problem: The work function for potassium is 2.20eV. What is the stopping potential for light of wavelength 400nm?

Solution:
$$eV_0 = hf - \varphi = hc / \lambda - \varphi = 4.136 \times 10^{-15} \times 3 \times 10^8 / (400 \times 10^{-9}) - 2.20 = 3.10 - 2.20 = 0.90eV$$
Thus stopping potential $V_0 = 0.90V$

Skill 23.2   Analyze bright-line spectra in terms of electron transitions

Quantum #	Radius
$n \to \infty$	$r_\infty \to \infty$
$\vdots$	$\vdots$
$n = 5$ ——	$r_5 = 25a_0$
$n = 4$ ——	$r_4 = 16a_0$
$n = 3$ ——	$r_3 = 9a_0$
$n = 2$ ——	$r_2 = 4a_0$
$n = 1$ ——	$r_1 = a_0$
$\oplus$ (H nucleus)	

An electron may exist at distinct radial distances (r_n) from the nucleus. These distances are proportional to the square of the **principal quantum number**, n. For a hydrogen atom (shown at left), the proportionality constant is called the **Bohr radius** ($a_0 = 5.29 \times 10^{-11}$ m). This value is the mean distance of an electron from the nucleus at the ground state of $n = 1$. The distances of other electron shells are found by the formula:
$$r_n = a_0 n^2.$$
As $n \to \infty$, the electron is no longer part of the hydrogen atom. Ionization occurs and the atom become an H^+ ion.

A quantum of energy (ΔE) emitted from or absorbed by an electron transition is directly proportional to the frequency of radiation. The proportionality constant between them is **Planck's constant** ($h = 6.63 \times 10^{-34}$ J·s):
$$\Delta E = h\nu \quad \text{and} \quad \Delta E = \frac{hc}{\lambda}.$$

The energy of an electron (E_n) is inversely proportional to its radius from the nucleus. For a hydrogen atom, only the principle quantum number determines the energy of an electron by the **Rydberg constant** ($R_H = 2.18 \times 10^{-18}$ J):
$$E_n = -\frac{R_H}{n^2}.$$

The Rydberg constant is used to determine the energy of a photon emitted or absorbed by an electron transition from one shell to another in the H atom:

$$\Delta E = R_H \left(\frac{1}{n_{initial}^2} - \frac{1}{n_{final}^2} \right).$$

When a photon is absorbed, n_{final} is greater than $n_{initial}$, resulting in positive values corresponding to an endothermic process. Ionization occurs when sufficient energy is added for the atom to lose its electron from the ground state. This corresponds to an electron transition from $n_{initial} = 1$ to $n_{final} \rightarrow \infty$. The Rydberg constant is the energy required to ionize one atom of hydrogen. Photon emission causes negative values corresponding to an exothermic process because $n_{initial}$ is greater than n_{final}.

Planck's constant and the speed of light are often used to express the Rydberg constant in units of s^{-1} or length. The formulas below determine the photon frequency or wavelength corresponding to a given electron transition:

$$v_{photon} = \left(\frac{R_H}{h} \right) \left| \frac{1}{n_{initial}^2} - \frac{1}{n_{final}^2} \right| \quad \text{and} \quad \lambda_{photon} = \frac{1}{\left(\frac{R_H}{hc} \right) \left| \frac{1}{n_{initial}^2} - \frac{1}{n_{final}^2} \right|}.$$

These formulas relate observed lines in the hydrogen spectrum to individual transitions from one quantum state to another.

A simple optical spectroscope separates visible light into distinct wavelengths by passing the light through a prism or diffraction grating. When electrons in hydrogen gas are excited inside a discharge tube, the emission spectroscope shown below detects photons at four visible wavelengths.

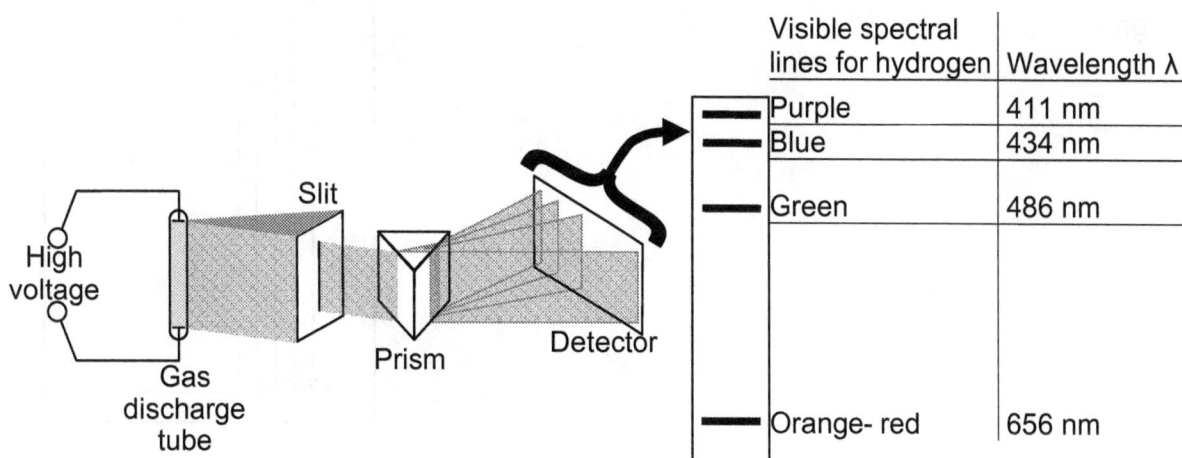

Visible spectral lines for hydrogen	Wavelength λ
Purple	411 nm
Blue	434 nm
Green	486 nm
Orange- red	656 nm

Every line in the hydrogen spectrum corresponds to a transition between electron energy levels. The four spectral lines from hydrogen emission spectroscopy in the visible range correspond to electron transitions from n = 3, 4, 5, and 6 to n =2 as shown in the table below.

Radiation type	Wavelength ×(nm)	Frequency ×(s^{-1})	Energy change ×E (J)	Electron transition $n_{initial} \rightarrow n_{final}$
Ultraviolet	$\leq$397	$\geq 7.55 \times 10^{14}$	$\leq -5.00 \times 10^{-19}$	$\infty \rightarrow 1, \ldots 2 \rightarrow 1$ $\infty \rightarrow 2, \ldots 7 \rightarrow 2$
Purple	411	7.31×10^{14}	-4.84×10^{-19}	$6 \rightarrow 2$
Blue	434	6.90×10^{14}	-4.58×10^{-19}	$5 \rightarrow 2$
Green	486	6.17×10^{14}	-4.09×10^{-19}	$4 \rightarrow 2$
Orange-red	656	4.57×10^{14}	-3.03×10^{-19}	$3 \rightarrow 2$
Infrared and beyond	$\geq$821	$\leq 3.65 \times 10^{14}$	$\geq -2.42 \times 10^{-19}$	$\infty \rightarrow 3, \ldots 4 \rightarrow 3$ $\infty \rightarrow 4, \ldots 5 \rightarrow 4$ $\vdots$

Most lines in the hydrogen spectrum are not at visible wavelengths. Larger energy transitions produce ultraviolet radiation and smaller energy transitions produce infrared or longer wavelengths of radiation. Transitions between the first three and the first six energy levels of the hydrogen atom are shown in the diagram to the right. The energy transitions producing the four visible spectral lines are colored grey.

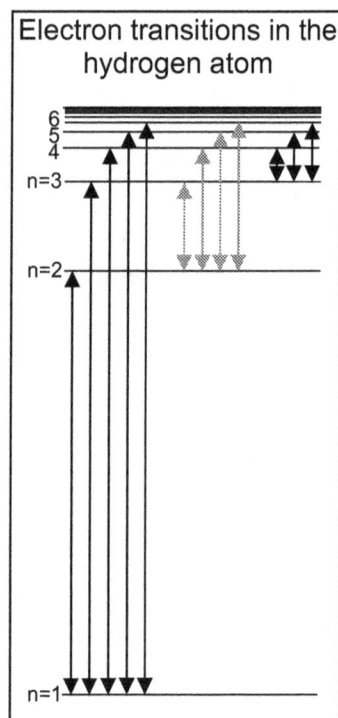

Electron transitions in the hydrogen atom

Skill 23.3   Understand the principles of stimulated emission of radiation as applied to lasers and masers

The names "laser" and "maser" are acronyms for "light amplification by stimulated emission of radiation" and "microwave amplification by stimulated emission of radiation." Thus, as is evident by their names, these two devices are based on the principle of stimulated emission.

According to quantum theory, an atom in an excited state has a certain probability in a given time frame for relaxing to a lower state through, for example, the emission of a photon of the same energy as the energy difference between the two states. Stimulated emission, however, may take place when the excited atom is perturbed by a passing photon of the same energy as the excitation. In such a case, the atom relaxes to a lower energy by emitting a new photon, resulting in a total of two photons of equal frequency (and, thus, energy) and equal phase. That is to say, the photons are coherent.

In order to produce a significant level of stimulated emission for application in typical lasers and masers, population inversion is required. Population inversion is a situation in which there are more atoms in a particular excited state than in a particular lower-energy state. In order to achieve this condition, the material must be "pumped," which can be performed using electromagnetic fields (light). Pumping to produce population inversion often requires three or more atomic energy levels.

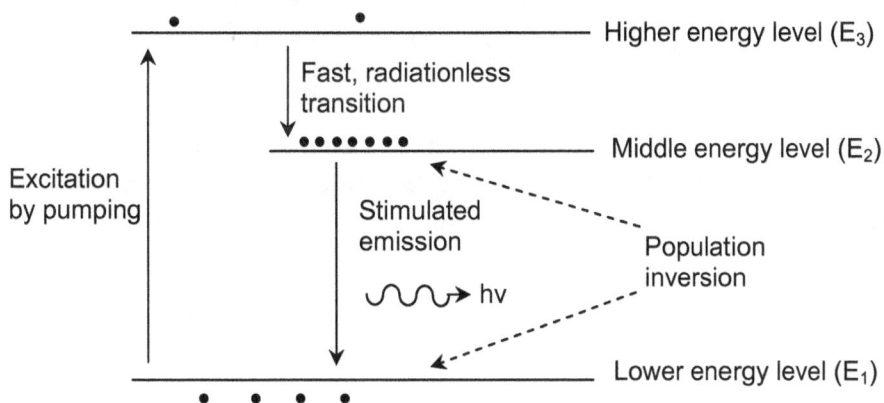

The above energy level diagram is for a so-called three-level laser (or maser). Other lasers may use more levels, depending on the material that is being pumped and the desired frequency output. The transition between the energy levels E_3 and E_2 is noted as being fast and radiationless; that is, almost as soon as the electron(s) of an atom are pumped to E_3, they decay to E_2 without radiating light. The energy difference can be emitted in the form of a phonon (vibration mode in the material, or heat).

The abundance of excited atoms due to population inversion allow for a "chain reaction" to form when photons of the proper frequency (proportional by Planck's constant to the energy difference between the excited state and lower-energy state) are incident. This results in a cascading increase in photon intensity through stimulated emission, thus producing the coherent, high-power light that is used by the laser or maser. This process has its limits, of course, as determined by how quickly the atoms can be pumped in relation to the relaxation processes. A saturation level exists beyond which the rate of stimulated emission cannot be increased. Spontaneously emitted photons can start the cascading process of stimulated emission.

The main components of a laser are a gain medium, an energy source (pump) for the gain medium and a resonant optical cavity with a partial transparency in one mirror. The resonant cavity is tuned to a particular frequency such that the photons of the desired laser frequency are coherent and, largely, isolated within the cavity. A partially transparent mirror on one end of the cavity allows some of the light to exit, thus producing the laser beam.

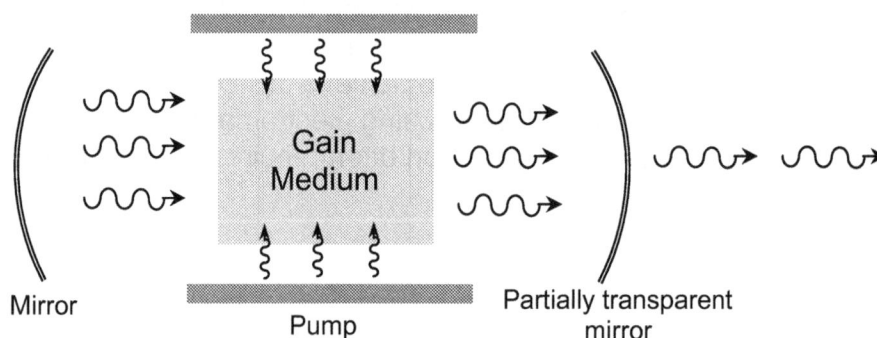

Masers, which operate at lower frequencies than lasers, generally rely on the same principles as lasers, although the types of gain media and resonant cavities may differ.

Skill 23.4   Analyze evidence supporting the dual nature of light and matter

Scientists have argued for years whether light is a wave or a stream of particles. Actually, light exhibits the behaviors of both waves and particles.

Wave characteristics

Light undergoes reflection, refraction, and diffraction just as any wave would. The image you see in a mirrored surface is the result of the reflection of the light waves off the surface. Light waves follow the "law of reflection," i.e. the angle at which the light wave approaches a flat reflecting surface is equal to the angle at which it leaves the surface. When light crosses the boundary between two different media, its path is bent, or refracted. Diffraction occurs when light encounters an obstacle in its path or passes through an opening. Light diffracts around the sides of an object causing the shadow of the object to appear fuzzy.

Another phenomenon unique to waves is wave interference. This characteristic describes what happens when two waves meet while traveling along the same medium. If light constructively interferes (trough meets trough or crest meets crest), the two light waves reinforce one another to produce a stronger light wave. However, if light destructively interferes (crest meets trough), the two light waves destroy each other and no light wave is produced.

Polarization changes unpolarized light into polarized light. This process can only occur with a transverse wave. An everyday example of polarization is found in polarized sunglasses which reduce glare.

Particle characteristics

Einstein came up with the quantum theory of light that states that light is made up of photons or quanta, discrete particles of electromagnetic radiation. The photons, or individual particles of light, have been shown to have isolated arrival times. A movie was taken of the comet Hyakutake that showed a breakdown of the photons traveling with the comet and scattered throughout the region. Some phenomena such as blackbody radiation and the photoelectric effect can only be explained using the particle nature of light.

Presently, a combination of the two theories, or wave - particle duality, is accepted.

COMPETENCY 24.0   Understand physical models of atomic structure and the nature of elementary particles

Skill 24.1   Analyze historic and contemporary models of atomic

In the West, the Greek philosophers Democritus and Leucippus first suggested the concept of the atom. They believed that all atoms were made of the same material but that varied sizes and shapes of atoms resulted in the varied properties of different materials. By the 19th century, John Dalton had advanced a theory stating that each element possesses atoms of a unique type. These atoms were also thought to be the smallest pieces of matter which could not be split or destroyed.

Atomic structure began to be better understood when, in 1897, JJ Thompson discovered the electron while working with cathode ray tubes. Thompson realized the negatively charged electrons were subatomic particles and formulated the "plum pudding model" of the atom to explain how the atom could still have a neutral charge overall. In this model, the negatively charged electrons were randomly present and free to move within a soup or cloud of positive charge.   Thompson likened this to the dried fruit that is distributed within the English dessert plum pudding though the electrons were free to move in his model. Ernest Rutherford disproved this model with the discovery of the nucleus in 1909. Rutherford proposed a new "planetary" model of the atom in which electrons orbited around a positively charged nucleus like planets around the sun. Over the next 20 years, protons and neutrons (subnuclear particles) were discovered while additional experiments showed the inadequacy of the planetary model.

As quantum theory was developed and popularized (primarily by Max Planck and Albert Einstein), chemists and physicists began to consider how it might apply to atomic structure. Niels Bohr put forward a model of the atom in which electrons could only orbit the nucleus in circular orbitals with specific distances from the nucleus, energy levels, and angular momentums. In this model, electrons could only make instantaneous "quantum leaps" between the fixed energy levels of the various orbitals. The Bohr model of the atom was altered slightly by Arnold Sommerfeld in 1916 to reflect the fact that the orbitals were elliptical instead of round.

Though the Bohr model is still thought to be largely correct, it was discovered that electrons do not truly occupy neat, cleanly defined orbitals. Rather, they exist as more of an "electron cloud." The work of Louis de Broglie, Erwin Schrödinger, and Werner Heisenberg showed that an electron can actually be located at any distance from the nucleus. However, we can find the *probability* that the electrons exists at given energy levels (i.e., in particular orbitals) and those probabilities will show that the electrons are most frequently organized within the orbitals originally described in the Bohr model.

Skill 24.2 Interpret notation used to represent elements, molecules, ions, and isotopes

Every element in the periodic table is represented by its own symbol consisting of one or more letters. (E.g. H for hydrogen, He for Helium, O for oxygen). In general, a molecule of an element or a compound is represented by a formula that shows what atoms and how many of each constitute the molecule. For instance, water consists of two hydrogen and one oxygen atom. Thus a molecule of water is represented as $H_2 0$. This is the kind of notation used in chemical equations.

Atoms and ions of a given element that differ in number of neutrons have a different mass and are called isotopes. In writing nuclear equations, where isotopes of the same element must be distinguished, it is helpful to use a notation where the number of nucleons and protons/electrons is included along with the element symbol. The identity of an element depends on the number of protons in the nucleus of the atom. This value is called the atomic number and it is sometimes written as a subscript before the symbol for the corresponding element. A nucleus with a specified number of protons and neutrons is called a nuclide, and a nuclear particle, either a proton or neutron, may be called a nucleon. The total number of nucleons is called the mass number and may be written as a superscript before the atomic symbol.

$$^{14}_{6}C$$ represents an atom of carbon with 6 protons and 8 neutrons.

The number of neutrons may be found by subtracting the atomic number from the mass number. For example, uranium- 235 has 235–92=143 neutrons because it has 235 nucleons and 92 protons.

An ion is an atom or molecule with a net positive or negative charge (due to an unequal number of protons and electrons) and is represented with a plus or minus sign and the number of excess electrons or protons placed at the top right-hand corner. For example, a positively charged sodium ion with one extra proton may be represented as Na^{+1}.

Skill 24.3 **Understand the relationship between the design of particle accelerators and elementary particle characteristics**

Elementary (fundamental) particles are, in many ways, theoretical. These particles are not seen with the unaided (or aided) human eye, and therefore must be understood based on other phenomena that are, presumably, associated with or caused by their presence or action. As such, experimentally investigating fundamental particles requires a theoretical framework for design of the experimental apparatus and for interpretation of the results. With the recent advent of inexpensive computing power, complex and repetitive calculations can be conducted quickly and relatively inexpensively, but the numbers that are output by the computer are only as good as the data that is input and the interpretation that is later applied to the results.

First, it is critical to note that the "resolution" of an experiment involving fundamental particles corresponds very closely to the kinetic energy of the particles involved. As with light (photons), the wavelength of the illuminating particles must be smaller than the smallest desired dimension of resolution in order to "see" the object of interest. As such, a particle accelerator must be designed with the dimensions of the proposed investigation clearly in mind. That is to say, the energy capabilities of the accelerator must meet or exceed the requirements for a given resolution (such as, for example, investigation of quarks).

The method for accelerating the particles is key; fundamental particles, of course, cannot be accelerated reliably by mechanical or conventional means. The use of electric and magnetic fields and waves allows for the acceleration of particles, but only charged particles. Thus, if neutral particles are to be investigated in this manner, their production must be possible using charged particles (e.g., by way of charged particle collisions). The characteristics of the particles to be investigated using the accelerator must then be considered carefully and integrated into the design of the equipment.

Some form of target must be used to facilitate the production of collision events and to allow study of the particles. Using the analogy of light shining on an object, fundamental particles are used to "illuminate" an object (perhaps another fundamental particle) to examine its behavior and structure. The target may simply consist of another beam of particles traveling in the opposite direction, forcing the particles to collide. Alternatively, some fixed target may be used and may contain any number of materials, depending on the experiments of interest.

Last, detection and transduction of data must be achieved through appropriate design of detectors and other equipment. This equipment is responsible for converting an event associated with the detection of a particle into a signal recognizable by a computer and, ultimately, analyzable by the experimenters. Of interest in the design process is, for example, the charge of the particles that originate from a collision event. These particles can be deflected by electric or magnetic fields, and the associated detection equipment can take advantage of these properties. Photons, an example of an uncharged particle, can be detected with photomultiplier tubes. Thus, the types of detection equipment necessary in the design of an accelerator depend on the characteristics of the particles of interest.

The four main points in the design of an accelerator include the particle kinetic energy (and, concomitantly, the resolution) necessary for the proposed experiments, the form of particle accelerator appropriate to the particular particles, the target to be used for producing collision events and the detection equipment necessary for the specific particles being investigated. Of course, proper computer handling of the data is likewise crucial. Since computers simply output numbers, a proper theory for interpretation is necessary.

COMPETENCY 25.0 **Understand the principles of radioactivity and types and characteristics of radiation, and analyze the process of radioactive decay and detection**

Skill 25.1 **Apply principles of the conservation of mass number and charge to balance equations for nuclear reactions.**

In alpha decay, an atom emits an alpha particle. An alpha particle contains two protons and two neutrons. This makes it identical to a helium nucleus and so an alpha particle may be written as He^{2+} or it can be denoted using the Greek letter α. Because a nucleus decaying through alpha radiation loses protons and neutrons, the mass of the atom loses about 4 Daltons and it actually becomes a different element (transmutation). For instance:

$$^{238}U \rightarrow {}^{234}Th + \alpha$$

This isotope of uranium weighs 238 Daltons. Because it is uranium, it has 92 protons, meaning it must have 146 neutrons. When it undergoes alpha decay it loses 2 protons and 2 neutrons. The alpha particle weighs 4 Daltons and the nucleus that has undergone decay weighs 234 Daltons. Thus mass is conserved. The decayed nucleus will have a charge reduced by that of 2 protons following the decay. However, the emitted alpha particle carries the additional charge due to its 2 protons. Thus both charge and mass are conserved over all.

Like alpha decay, beta decay involves emission of a particle. In this case, though, it is a beta particle, which is either an electron or positron. Note that a positron is the antimatter equivalent of an electron and so these particles are often denoted β^- and β^+. Beta plus and minus decay occur via roughly opposite paths. In beta minus decay, a neutron is converted in a proton (specifically, a down quark is converted to an up quark), an electron and an anti-neutrino; the latter two are emitted. In beta plus decay, on the other hand, a proton is converted to a neutron, a positron, and a neutrino; again, the latter two are emitted. As in alpha decay, a nucleus undergoing beta decay is transmuted into a different element because the number of protons is altered. However, because the total number of nucleons remains unchanged, the atomic mass remains the same (note, that the neutron is actually slightly heavier than a proton so mass is gained during beta plus decay).

An examples of beta minus decay is as follows:

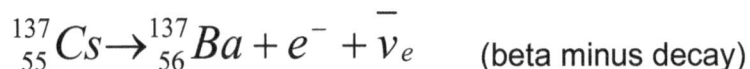

$$^{137}_{55}Cs \rightarrow {}^{137}_{56}Ba + e^- + \bar{v}_e \quad \text{(beta minus decay)}$$

This isotope of caesium weighs 137 Daltons and, like all caesium isotopes, it has 55 protons. When it undergoes beta minus decay, a neutron is converted to a proton and an electron and an anti-neutrino are lost.

The total mass-energy of the system is conserved since the difference in mass between a neutron and an electron plus proton is balanced by the energy of the emitted electron and the anti-neutrino. In beta minus decay a neutron, with no charge, is split into a positively charged proton and a negatively charged electron. Thus the conservation of charge is satisfied. The electron is emitted, while the proton remains in the nucleus. With one extra proton, the nucleus is now a barium isotope.

In order to conserve mass-energy of the system, beta plus decay cannot occur in isolation but only in a nucleus since the mass of a neutron is greater than the mass of a proton plus electron. The difference in binding energy of the mother and daughter nucleus provides the additional energy needed for the reaction to go through. In the example below, charge is conserved when a positively charged proton is converted into a positively charged positron. With one fewer proton, the decayed nucleus is now a neon isotope weighing 22 Daltons.

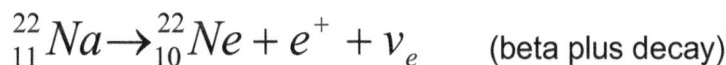

$$\,^{22}_{11}Na \rightarrow \,^{22}_{10}Ne + e^+ + v_e \qquad \text{(beta plus decay)}$$

Problems involving balancing nuclear equations can involve simple radioactive decay, fission, fusion, and other nuclear processes. In all cases, both mass-energy and charge must be balanced.

Problem: Uranium 235 is used as a nuclear fuel in a chain reaction. The reaction is initiated by a single neutron and produces barium 141, an unknown isotope, and 3 neutrons that can go on to propagate the chain reaction. Determine the unknown isotope. Assume that the kinetic energies and the energy released in the reaction is negligible compared to the masses of the isotopes produced.

Solution: We can begin by writing out the reaction, leaving open the unknown isotope X.

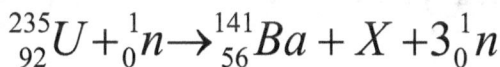

$$\,^{235}_{92}U + \,^{1}_{0}n \rightarrow \,^{141}_{56}Ba + X + 3\,^{1}_{0}n$$

We begin with the charge balance; since the neutron has no charge, the unknown isotope must have 36 protons. Consulting a periodic table, we see that this will mean it is a Krypton isotope. Now we can balance the mass. Since the original nucleus weighed 235 Daltons and one neutron was added to it, the total mass of the resultant nuclei must be 236. So, we can simple subtract the weight of the barium isotope and the 3 new neutrons to find the unknown weight:

$$236-141-3=92$$

Thus our unknown isotope is krypton 92, making the balanced equation:

$$\,^{235}_{92}U + \,^{1}_{0}n \rightarrow \,^{141}_{56}Ba + \,^{92}_{36}Kr + 3\,^{1}_{0}n$$

Skill 25.2   Analyze radioactive decay in terms of the half-life concept

While the radioactive decay of an individual atom is impossible to predict, a mass of radioactive material will decay at a specific rate. Radioactive isotopes exhibit exponential decay and we can express this decay in a useful equation as follows:

$$A=A_0e^{kt}$$

Where A is the amount of radioactive material remaining after time t, A_0 is the original amount of radioactive material, t is the elapsed time, and k is the unique activity of the radioactive material. Note that k is unique to each radioactive isotope and it specifies how quickly the material decays. Sometimes it is convenient to express the rate of decay as half-life. A half-life is the time needed for half a given mass of radioactive material to decay. Thus, after one half-life, 50% of an original mass will have decayed, after two half lives, 75% will have decayed and so on.

Let's examine a sample problem related to radioactive decay.

Problem: Radiocarbon dating has been used extensively to determine the age of fossilized organic remains. It is based on the fact that while most of the carbon atoms in living things is ^{12}C, a small percentage is ^{14}C. Since ^{14}C is a radioactive isotope, it is lost from a fossilized specimen at a specific rate following the death of an organism. The original and current mass of ^{14}C can be inferred from the relative amount of ^{12}C. So, if the half-life of ^{14}C is 5730 years and a specimen that originally contained 1.28 mg of ^{14}C now contains 0.10 mg, how old is the specimen?

In certain problems, we may be simply provided with the activity, k, but in this problem we must use the information given about half-life to solve for k.

Since we know that after one half-life, 50% of the material remains radioactive, we can plug into the governing equation above:

$$A=A_0e^{kt}$$

$$0.5\,A_0=A_0e^{5730k}$$

$$k=(\ln(0.5))/5730=-0.0001209$$

Having determined k, we can use this same equation again to determine how old the specimen described above must be:

$$A = A_0 e^{kt}$$

$$0.10 = 1.28 e^{-0.0001209t}$$

$$t = \frac{\ln\left(\dfrac{0.10}{1.28}\right)}{-0.0001209} = 21087$$

Thus, the specimen is 21,087 years old.

Note that this same equation can be used to calculate the half-life of an isotope if information regarding the decay after a given number of years were provided.

A radioactive isotope often decays into another element that is radioactive as well and continues to decay into a third element. The process continues until a stable element is reached. This chain of disintegration is known as a **decay chain** or a **nuclear disintegration series**. For example, Uranium 238 decays into Radium 226 which decays further into Radon 222.

Skill 25.3   Analyze the nuclear disintegration series for a given isotope

Radioactive elements usually decay to a stable product through a progression of intermediate radionuclides.   This chain of decay products occurs through processes such as alpha (α) and beta (β⁻) decay.

There are three main decay chains known as the Thorium, Radium, and Actinium series. The respective starting isotopes of these series, ^{232}Th, ^{238}U, and ^{235}U, have been decaying since the formation of the earth. Each decay chain ends with a different, but stable isotope of lead. The Thorium decay series is shown as an example below. The initial radionuclide is ^{232}Th and the stable decay product is ^{208}Pb. The decay mechanism for each intermediate product is shown above the arrows.

Thorium Decay Series

$$\overset{\alpha}{\longrightarrow}\ \overset{\beta^-}{\longrightarrow}\ \overset{\beta^-}{\longrightarrow}\ \overset{\alpha}{\longrightarrow}\ \overset{\alpha}{\longrightarrow}\ \overset{\alpha}{\longrightarrow}\ \overset{\alpha}{\longrightarrow}\ \overset{\beta^-}{\longrightarrow}\ \overset{\beta^-}{\longrightarrow}\ \overset{\alpha}{\longrightarrow}$$

^{232}Th → ^{228}Ra → ^{228}Ac → ^{228}Th → ^{224}Ra → ^{220}Rn → ^{216}Po → ^{212}Pb → ^{212}Bi

$$\begin{array}{ccc} & \overset{\beta^-}{\longrightarrow} & ^{212}\text{Po} \overset{\alpha}{\longrightarrow} \\ & \overset{\alpha}{\longrightarrow} & ^{208}\text{Tl} \overset{\beta^-}{\longrightarrow} \end{array} \quad ^{208}\text{Pb}$$

Decay chains are often discussed in terms parent and daughter isotopes. A parent isotope undergoes decay to a daughter isotope. The decay of anyone particular atom is considered spontaneous. However, a larger sample of radionuclides decays exponentially. The exact rate at which decay occurs is determined by the half-life of the radionuclide.

It is important to understand that parent isotopes do not always decay in ways that decrease both the mass and atomic number. For example, an unstable 232-Thorium atom decays first to 228-Radium via alpha decay. Therefore, the parent isotope decreases in mass number by four and in atomic (proton) number by two. In the second step of the Thorium decay series, a 228-Radium atom can be transformed to 228-Actinium atom via a Beta minus decay. In such a decay step, the mass number will remain the same and the atomic number actually increases by one. Thus, the atomic number progression for the first two steps of the Thorium decay series would be 232-Thorium with 90 protons to 228-Radium with 88 protons to 228-Actinium with 89 protons. When analyzing a particular decay series, the type of decay determines the daughter isotope product.

While almost all decay chains lead to a stable isotope lead, shorter decay chains occur as a result of isotopes formed by cosmic radiation approaching the earth.

Skill 25.4   Understand the basic operation of types of radiation detectors

Radiation detectors have been designed for a number of different applications and rely on various mechanisms for converting radiation into an electrical signal. Perhaps the most well-known form of radiation detector is the **Geiger counter**, a device based on the Geiger-Müller (GM) tube. The GM tube is an isolated chamber, usually filled with a combination of an inert gas, such as helium or neon, and a halogen gas, such as chlorine. Inside the tube, an anode and a cathode are present with a strong electric potential difference of hundreds of volts. When ionizing radiation enters the tube, some of the gas molecules are ionized with the positive ion being attracted towards the negatively charged anode and the negative ion (an electron) being attracted towards the positively charged cathode. The strong electric potential causes the ions to gain sufficient kinetic energy to further ionize other gas particles resulting in a cascade of particles and a detectable current. Depending on the specific materials used for the Geiger counter (e.g., the detection "window"), the device can detect alpha particles (helium nuclei), beta particles (an electron or positron resulting from a nuclear beta decay) or gamma rays (high-energy photons). So-called proportional counters, which operate on similar principles, are able to determine the energy of a particle of ionizing radiation through counting the "avalanches" that occur when gas molecules in the tube are ionized and accelerated to cause further ionization.

Photomultiplier tubes are an often-used method for detection of electromagnetic radiation in the visible region of the spectrum as well as in portions of the infrared and ultraviolet regions. The photoelectric effect is the driving phenomenon behind the operation of these devices. Photons that strike a so-called photocathode can cause the emission of an electron which is then accelerated towards a positively charged electrode called a dynode. A number of progressively more positively charged dynodes are arranged in the photomultiplier tube to cause a cascading effect. Each dynode produces further electrons in response to the impact of the incident electron, then "focuses" the emitted electrons onto the next dynode. Charge is collected at an anode, resulting in an electrical signal. The signal current produced by a photomultiplier tube is proportional to the intensity of the incident light.

Another device for detecting ionizing radiation is the **scintillation counter**. In this case a so-called scintillator is used. This is a material that produces a burst of light as a result of incident high-energy particles (such as photons). The incident high-energy particle loses some of its energy in the scintillator and the energy is converted into a number of lower-energy photons. A photomultiplier tube is then used to measure the intensity of the light pulse which yields a measure of the energy of the incident particle. Scintillation counters are used in such contexts as high-energy physics for detection equipment in particle accelerators.

Another radiation detector used in particle physics is the **calorimeter (or microcalorimeter)**. High-energy particles, rather than producing light, as with the scintillator, lose some of their energy to produce heat through incidence on the material used in the calorimeter. A thermometer is used to detect the change in the temperature of the material with the signal intensity providing information on the energy of the particle.

A number of other devices can be used in various contexts involving detection of radiation. Particle track detectors, such as bubble chambers or other devices containing solid, liquid or gaseous materials can be used to identify the passage of charged (or sometimes uncharged) particles. Trails left in the material due to incidence of various particles, which are influenced by a magnetic field, can be analyzed to determine the energy, charge and other features of the particles.

COMPETENCY 26.0   Understand types and characteristics of nuclear reactions, methods of initiating and controlling them, and applications of nuclear reactions to the generation of electricity

Skill 26.1   Analyze characteristics of fission and fusion reactions

Nuclear fission involves the splitting of an atomic nucleus with a subsequent emission of energy in the form of photons with a specific frequency or other particles with a specific kinetic energy. As with combustion, the energy released from a fission reaction can be used to power turbines and to generate electricity. About 20% of the electrical power generated in the United States comes from nuclear reactors.

The heat produced from the controlled nuclear reaction can be used in the same manner as heat from combustion to produce steam and turn a turbine, thus generating electricity. The hyperboloid towers typically seen at nuclear power plants serve the purpose of cooling.

Fusion and fission are two complementary reactions with fusion being the preferred reaction for elements lighter than iron (atomic number 26) or nickel (atomic number 28) and fission being the preferred reaction for elements heavier than iron or nickel. In the case of fusion, energy is released for the lighter elements but is absorbed for the heavier elements. Fission operates in the reverse manner.

Fusion reactions and binding energy (mass defect)

The energy released during a fusion reaction is the binding energy which corresponds to a conversion of some of the mass in the constituent particles into emitted energy (mass defect). The mass defect allows the fused nucleus to exist at a lower energy level, thus making it more stable. Although the energy required to fuse two light nuclei is considerable, the energy released by the reaction is even greater and, under the right circumstances, allows for a self-sustaining (chain) reaction. Such a self-sustaining reaction requires that the energy needed to bring two nuclei to within sufficient proximity for fusion to take place is less than the energy liberated when the nuclei are fused. This initial required energy is necessary to overcome the electrostatic repulsion between the positively charged nuclei.

Fusion is believed to be the primary source of solar energy. In a more terrestrial context, fusion has been exploited for the purpose of creating the so-called hydrogen bomb, which involves an uncontrolled chain reaction that releases an immense amount of destructive energy. In the context of the Sun, fusion provides the energy that is a major factor in maintaining the habitability of the Earth.

Stellar fusion

Since fusion takes place in stars, it is a process that is central to the study of astronomy. Two important reactions involved in stellar fusion are the so-called proton-proton (PP) chain and carbon-nitrogen-oxygen (CNO) cycle. The PP chain involves a process whereby protons are converted into heavier helium nuclei with the concomitant release of energy. It is the primary engine of energy production in stars similar to the Sun. The CNO cycle involves interconversion of carbon, nitrogen and oxygen nuclei and is the driving process in larger stars. Although the PP fusion process takes place through a number of stages, the basic components and products of the reaction are described below, where protons (^{1}H) combine to form a helium nucleus (^{4}He), two neutrinos (v), two positrons (e^+) and energy in the form of photons.

Skill 26.2   Understand the operation of components of a nuclear reactor

In order to produce a chain reaction, rods filled with **nuclear fuel pellets** (such as ^{235}U) are used inside the reactor core. Since fission requires relatively slow-moving neutrons, a so-called **moderator** is needed to slow down the fast neutrons produced from fission elsewhere in the core, thus making the chain reaction possible. The moderator can be in the form of water (standard water or deuterium-based "heavy water"), which can also serve as a coolant. While these steps are necessary to sustain a nuclear reaction, it is also equally important to control the reaction. This is accomplished through the use of **control rods** which are filled with a neutron-absorbing material such as boron or silver. While these materials capture free neutrons, they are not themselves fissile and thus their presence limits the nuclear reaction in the core. The location of the control rods can be adjusted for dynamic control of the reaction. The entire apparatus of fuel rods, control rods and moderator (and coolant) is contained inside a pressure vessel.

Skill 26.3   Apply the principle of conservation of mass-energy to calculate nuclear mass defect and binding energy

Independently of one another, mass and energy are not necessarily conserved, as one may be converted to another in certain instances. Together, however, as mass-energy, they are indeed conserved. In most macroscopic circumstances, mass and energy are each conserved individually. On a microscopic scale, however, especially in the realm of particle physics, interconversion of mass and energy can take place to a significant extent. In these cases, the total conservation of mass-energy in a closed system is the most general conservation law.

Binding energy and nuclear mass defect
The binding energy and nuclear mass defect are two complementary concepts associated with the interaction of fundamental particles. The protons that form the nucleus of an atom, for example, are positively charged and, thus, have a mutual electrostatic repulsion. As a result, the energy of a system of closely packed protons is high. Nevertheless, the atomic nucleus is still quite stable. This stability results from a release of energy upon formation of the nucleus, thus lowering the total energy of the system. The released energy is called the binding energy. Conversely, for the reverse process, the binding energy is the energy required to split the nucleus into its component particles, and is often described generally as an energy per nucleon.

The source of the released binding energy during a fusion of protons (or protons and neutrons) into a nucleus is a portion of the mass of the system. According to special relativity, the mass of an object (whether rest or relativistic mass) has an equivalent energy ($E = mc^2$ in the case of rest mass). In the context of nuclear physics, protons that are fused into a nucleus are able to release energy at the expense of a portion of their mass. This nuclear mass defect is the emission of a certain portion of the mass of the constituent particles, in the form of energy, to lower the total energy of the whole (e.g., the nucleus). Quantum chromodynamics (QCD) provides a more fundamental explanation of this phenomenon through the strong nuclear force, or color force. QCD details the interactions of the quarks that compose the substructure of protons and neutrons by way of the color charge, which is a property of quarks, and gluons, which are the mediating boson for the strong force.

Conservation of mass-energy
In light of these basic concepts surrounding the conservation of mass-energy, the nuclear mass defect and binding energy can be calculated using the relativistic relationship of mass and energy. The calculation can be performed by noting the difference in mass between the constituent particles and the final product of a reaction.

Problem: What is the binding energy of a deuteron (a nucleus composed of a proton and a neutron)?

Solution: The binding energy is the equivalent energy resulting from the difference in mass between the deuteron and the constituent particles.

Proton mass = 1.6726 x 10^{-27} kg
Neutron mass = 1.6749 x 10^{-27} kg
Deuteron mass = 3.3436 x 10^{-27} kg

Mass difference: $\Delta m = \left(1.6726 \times 10^{-27} + 1.6749 \times 10^{-27}\right) \text{kg} - 3.3436 \times 10^{-27} \text{kg}$
$\Delta m = 3.9 \times 10^{-30} \text{kg}$
Binding energy: $\Delta mc^2 = 3.9 \times 10^{-30} \text{kg} \cdot \left(2.9979 \times 10^8 \, \text{m/s}\right)^2 = 3.5051 \times 10^{-13} \text{J}$

Skill 26.4 **Analyze the reasons for using the isotopes commonly used to fuel nuclear reactors and the problems associated with the waste products generated by nuclear reactions**

If controlled, the neutron flux resulting from spontaneously decaying radioactive materials can be used to instigate further decay. With Uranium (^{235}U, specifically), for example, a neutron from another decay can cause the atom to split, thus releasing several neutrons in addition to energy. These neutrons can also cause further decay, resulting in a **chain reaction**. Nuclear weapons involve an uncontrolled chain reaction that yields a tremendously powerful explosion; nuclear power generators take advantage of the same effect, but in a controlled manner.

While nuclear power generation does not produce the same kind of carbon-based emissions as combustion of oil or coal, for example, there are other equally (if not more so) undesirable by-products. Waste products from the nuclear reaction are, themselves, often highly radioactive, posing an environmental threat. Ionizing radiation from such waste can cause tremendous damage to living organisms. Since there is no apparent way to easily neutralize these by-products, they must be stored securely to prevent contamination of the environment. This, of course, requires a storage facility and adequate isolation. Security and structural integrity of these facilities must be continually monitored.

Another issue revolving around the use of nuclear power is the ability to use a nuclear reactor to produce weapons-grade materials for use in explosive nuclear devices. This poses a tremendous threat since nuclear weapons are able to cause untold destruction as seen at the close of the Second World War in Japan. Furthermore, in the context of international terrorism, nuclear power plants in non-aggressive nations can still be a threat if attacked.

Thus, the plants themselves, although they may be operating for only peaceful purposes, could be damaged in such a way that nuclear materials are released into the environment, or, in a worst-case scenario, a runaway nuclear reaction is initiated. Even otherwise benign human error can lead to catastrophic consequences. Thus, in light of these considerations, a debate continues concerning nuclear power as the dangers and the benefits are weighed.

Sample Test

1. A projectile with a mass of 1.0 kg has a muzzle velocity of 1500.0 m/s when it is fired from a cannon with a mass of 500.0 kg. If the cannon slides on a frictionless track, it will recoil with a velocity of _____ m/s.

   A. 2.4

   B. 3.0

   C. 3.5

   D. 1500

2. The weight of an object on the earth's surface is designated x. When it is two earth's radii from the surface of the earth, its weight will be

   A. $x/4$

   B. $x/9$

   C. $4x$

   D. $16x$

3. If a force of magnitude F gives a mass M an acceleration A, then a force $3F$ would give a mass $3M$ an acceleration

   A. A

   B. $12A$

   C. $A/2$

   D. $6A$

4. A car (mass m_1) is driving at velocity v, when it smashes into an unmoving car (mass m_2), locking bumpers. Both cars move together at the same velocity. The common velocity will be given by

   A. m_1v/m_2

   B. m_2v/m_1

   C. $m_1v/(m_1 + m_2)$

   D. $(m_1 + m_2)v/m_1$

5. When acceleration is plotted versus time, the area under the graph represents

   A. Time

   B. Distance

   C. Velocity

   D. Acceleration

6. An inclined plane is tilted by gradually increasing the angle of elevation θ, until the block will slide down at a constant velocity. The coefficient of friction, μ_k, is given by

   A. cos θ

   B. sin θ

   C. cosecant θ

   D. tangent θ

7. Use the information on heats below to solve this problem. An ice block at 0° Celsius is dropped into 100 g of liquid water at 18° Celsius. When thermal equilibrium is achieved, only liquid water at 0° Celsius is left. What was the mass, in grams, of the original block of ice?

   Given: Heat of fusion of ice = 80 cal/g
   Heat of vaporization of ice = 540 cal/g
   Specific Heat of ice = 0.50 cal/g°C
   Specific Heat of water = 1 cal/g°C

   A. 2.0

   B. 5.0

   C. 10.0

   D. 22.5

8. The combination of overtones produced by a musical instrument is known as its

   A. Timbre

   B. Chromaticity

   C. Resonant Frequency

   D. Flatness

9. A long copper bar has a temperature of 60°C at one end and 0°C at the other. The bar reaches thermal equilibrium (barring outside influences) by the process of heat

   A. Fusion

   B. Convection

   C. Conduction

   D. Microwaving

10. The First Law of Thermodynamics takes the form dU = dW when the conditions are

   A. Isobaric

   B. Isochloremic

   C. Isothermal

   D. Adiabatic

11. Given a vase full of water, with holes punched at various heights. The water squirts out of the holes, achieving different distances before hitting the ground. Which of the following accurately describes the situation?

A. Water from higher holes goes farther, due to Pascal's Principle.

B. Water from higher holes goes farther, due to Bernoulli's Principle.

C. Water from lower holes goes farther, due to Pascal's Principle.

D. Water from lower holes goes farther, due to Bernoulli's Principle.

12. A stationary sound source produces a wave of frequency F. An observer at position A is moving toward the horn, while an observer at position B is moving away from the horn. Which of the following is true?

A. $F_A < F < F_B$

B. $F_B < F < F_A$

C. $F < F_A < F_B$

D. $F_B < F_A < F$

13. The electric force in Newtons, on two small objects (each charged to – 10 microCoulombs and separated by 2 meters) is

A. 1.0

B. 9.81

C. 31.0

D. 0.225

14. A 10 ohm resistor and a 50 ohm resistor are connected in parallel. If the current in the 10 ohm resistor is 5 amperes, the current (in amperes) running through the 50 ohm resistor is

A. 1

B. 50

C. 25

D. 60

15. Fahrenheit and Celsius thermometers have the same temperature reading at

A. 100 degrees

B. - 40 degrees

C. Absolute Zero

D. 40 degrees

16. Which of the following apparatus can be used to measure the wavelength of a sound produced by a tuning fork?

    A. A glass cylinder, some water, and iron filings

    B. A glass cylinder, a meter stick, and some water

    C. A metronome and some ice water

    D. A comb and some tissue

17. When the current flowing through a fixed resistance is doubled, the amount of heat generated is

    A. Quadrupled

    B. Doubled

    C. Multiplied by pi

    D. Halved

18. The current induced in a coil is defined by which of the following laws?

    A. Lenz's Law

    B. Burke's Law

    C. The Law of Spontaneous Combustion

    D. Snell's Law

19. If an object is 20 cm from a convex lens whose focal length is 10 cm, the image is:

    A. Virtual and upright

    B. Real and inverted

    C. Larger than the object

    D. Smaller than the object

20. A cooking thermometer in an oven works because the metals it is composed of have different

    A. Melting points

    B. Heat convection

    C. Magnetic fields

    D. Coefficients of expansion

21. In an experiment where a brass cylinder is transferred from boiling water into a beaker of cold water with a thermometer in it, we are measuring

    A. Fluid viscosity

    B. Heat of fission

    C. Specific heat

    D. Nonspecific heat

22. A temperature change of 40 degrees Celsius is equal to a change in Fahrenheit degrees of

    A. 40

    B. 20

    C. 72

    D. 112

23. The number of calories required to raise the temperature of 40 grams of water at 30°C to steam at 100°C is

    A. 7500

    B. 23,000

    C. 24,400

    D. 30,500

24. The boiling point of water on the Kelvin scale is closest to

    A. 112 K

    B. 212 K

    C. 373 K

    D. 473 K

25. The kinetic energy of an object is _____ proportional to its _____.

    A. Inversely...inertia

    B. Inversely...velocity

    C. Directly...mass

    D. Directly...time

26. A hollow conducting sphere of radius R is charged with a total charge Q. What is the magnitude of the electric field at a distance r (given r<R) from the center of the sphere? (k is the electrostatic constant)

    A. 0

    B. $k\,Q/R^2$

    C. $k\,Q/(R^2 - r^2)$

    D. $k\,Q/(R - r)^2$

27. A quantum of light energy is called a

    A. Dalton

    B. Photon

    C. Curie

    D. Heat Packet

28. The following statements about sound waves are true *except*

    A. Sound travels faster in liquids than in gases.

    B. Sound waves travel through a vacuum.

    C. Sound travels faster through solids than liquids.

    D. Ultrasound can be reflected by the human body.

29. The greatest number of 100 watt lamps that can be connected in parallel with a 120 volt system without blowing a 5 amp fuse is

    A. 24

    B. 12

    C. 6

    D. 1

30. A monochromatic ray of light passes from air to a thick slab of glass (n = 1.41) at an angle of 45° from the normal. At what angle does it leave the air/glass interface?

    A. 45°

    B. 30°

    C. 15°

    D. 55°

31. The magnitude of a force is

    A. Directly proportional to mass and inversely to acceleration

    B. Inversely proportional to mass and directly to acceleration

    C. Directly proportional to both mass and acceleration

    D. Inversely proportional to both mass and acceleration

32. A semi-conductor allows current to flow

    A. Never

    B. Always

    C. As long as it stays below a maximum temperature

    D. When a minimum voltage is applied

33. One reason to use salt for melting ice on roads in the winter is that

    A. Salt lowers the freezing point of water.

    B. Salt causes a foaming action, which increases traction.

    C. Salt is more readily available than sugar.

    D. Salt increases the conductivity of water.

34. Automobile mirrors that have a sign, "objects are closer than they appear" say so because

    A. The real image of an obstacle, through a converging lens, appears farther away than the object.

    B. The real or virtual image of an obstacle, through a converging mirror, appears farther away than the object.

    C. The real image of an obstacle, through a diverging lens, appears farther away than the object.

    D. The virtual image of an obstacle, through a diverging mirror, appears farther away than the object.

35. A gas maintained at a constant pressure has a specific heat which is greater than its specific heat when maintained at constant volume, because

    A. The Coefficient of Expansion changes.

    B. The gas enlarges as a whole.

    C. Brownian motion causes random activity.

    D. Work is done to expand the gas.

36. Consider the shear modulus of water, and that of mercury. Which of the following is true?

    A. Mercury's shear modulus indicates that it is the only choice of fluid for a thermometer.

    B. The shear modulus of each of these is zero.

    C. The shear modulus of mercury is higher than that of water.

    D. The shear modulus of water is higher than that of mercury.

37. A skateboarder accelerates down a ramp, with constant acceleration of two meters per second squared, from rest. The distance in meters, covered after four seconds, is

    A. 10

    B. 16

    C. 23

    D. 37

38. Which of the following units is not used to measure torque?

    A. slug ft

    B. lb ft

    C. N m

    D. dyne cm

39. In a nuclear pile, the control rods are composed of

    A. Boron

    B. Einsteinium

    C. Isoptocarpine

    D. Phlogiston

40. All of the following use semi-conductor technology, except a(n):

    A. Transistor

    B. Diode

    C. Capacitor

    D. Operational Amplifier

41. Ten grams of a sample of a radioactive material (half-life = 12 days) were stored for 48 days and re-weighed. The new mass of material was

    A. 1.25 g

    B. 2.5 g

    C. 0.83 g

    D. 0.625 g

42. When a radioactive material emits an alpha particle only, its atomic number will

    A. Decrease

    B. Increase

    C. Remain unchanged

    D. Change randomly

43. The sun's energy is produced primarily by

 A. Fission

 B. Explosion

 C. Combustion

 D. Fusion

44. A crew is on-board a spaceship, traveling at 60% of the speed of light with respect to the earth. The crew measures the length of their ship to be 240 meters. When a ground-based crew measures the apparent length of the ship, it equals

 A. 400 m

 B. 300 m

 C. 240 m

 D. 192 m

45. Which of the following pairs of elements are not found to fuse in the centers of stars?

 A. Oxygen and Helium

 B. Carbon and Hydrogen

 C. Beryllium and Helium

 D. Cobalt and Hydrogen

46. A calorie is the amount of heat energy that will

 A. Raise the temperature of one gram of water from 14.5° C to 15.5° C.

 B. Lower the temperature of one gram of water from 16.5° C to 15.5° C

 C. Raise the temperature of one gram of water from 32° F to 33° F

 D. Cause water to boil at two atmospheres of pressure.

47. Bohr's theory of the atom was the first to quantize

 A. Work

 B. Angular Momentum

 C. Torque

 D. Duality

48. A uniform pole weighing 100 grams, that is one meter in length, is supported by a pivot at 40 centimeters from the left end. In order to maintain static position, a 200 gram mass must be placed _____ centimeters from the left end.

 A. 10

 B. 45

 C. 35

 D. 50

49. A classroom demonstration shows a needle floating in a tray of water. This demonstrates the property of

   A. Specific Heat

   B. Surface Tension

   C. Oil-Water Interference

   D. Archimedes' Principle

50. Two neutral isotopes of a chemical element have the same numbers of

   A. Electrons and Neutrons

   B. Electrons and Protons

   C. Protons and Neutrons

   D. Electrons, Neutrons, and Protons

51. A mass is moving at constant speed in a circular path. Choose the true statement below:

   A. Two forces in equilibrium are acting on the mass.

   B. No forces are acting on the mass.

   C. One centripetal force is acting on the mass.

   D. One force tangent to the circle is acting on the mass.

52. A light bulb is connected in series with a rotating coil within a magnetic field. The brightness of the light may be increased by any of the following except:

   A. Rotating the coil more rapidly.

   B. Using more loops in the coil.

   C. Using a different color wire for the coil.

   D. Using a stronger magnetic field.

53. The use of two circuits next to each other, with a change in current in the primary circuit, demonstrates

   A. Mutual current induction

   B. Dielectric constancy

   C. Harmonic resonance

   D. Resistance variation

54. A brick and hammer fall from a ledge at the same time. They would be expected to

    A. Reach the ground at the same time

    B. Accelerate at different rates due to difference in weight

    C. Accelerate at different rates due to difference in potential energy

    D. Accelerate at different rates due to difference in kinetic energy

55. The potential difference across a five Ohm resistor is five Volts. The power used by the resistor, in Watts, is

    A. 1

    B. 5

    C. 10

    D. 20

56. An object two meters tall is speeding toward a plane mirror at 10 m/s. What happens to the image as it nears the surface of the mirror?

    A. It becomes inverted.

    B. The Doppler Effect must be considered.

    C. It remains two meters tall.

    D. It changes from a real image to a virtual image.

57. The highest energy is associated with

    A. UV radiation

    B. Yellow light

    C. Infrared radiation

    D. Gamma radiation

58. The constant of proportionality between the energy and the frequency of electromagnetic radiation is known as the

    A. Rydberg constant

    B. Energy constant

    C. Planck constant

    D. Einstein constant

59. A simple pendulum with a period of one second has its mass doubled. If the length of the string is quadrupled, the new period will be

    A. 1 second

    B. 2 seconds

    C. 3 seconds

    D. 5 seconds

60. A vibrating string's frequency is _____ proportional to the _____.

    A. Directly; Square root of the tension

    B. Inversely; Length of the string

    C. Inversely; Squared length of the string

    D. Inversely; Force of the plectrum

61. When an electron is "orbiting" the nucleus in an atom, it is said to possess an intrinsic spin (spin angular momentum). How many values can this spin have in any given electron?

    A. 1

    B. 2

    C. 3

    D. 8

62. Electrons are

    A. More massive than neutrons

    B. Positively charged

    C. Neutrally charged

    D. Negatively charged

63. Rainbows are created by

    A. Reflection, dispersion, and recombination

    B. Reflection, resistance, and expansion

    C. Reflection, compression, and specific heat

    D. Reflection, refraction, and dispersion

64. In order to switch between two different reference frames in special relativity, we use the _____ transformation.

    A. Galilean

    B. Lorentz

    C. Euclidean

    D. Laplace

65. A baseball is thrown with an initial velocity of 30 m/s at an angle of 45°. Neglecting air resistance, how far away will the ball land?

    A. 92 m

    B. 78 m

    C. 65 m

    D. 46 m

66. If one sound is ten decibels louder than another, the ratio of the intensity of the first to the second is

    A. 20:1

    B. 10:1

    C. 1:1

    D. 1:10

67. A wave has speed 60 m/s and wavelength 30,000 m. What is the frequency of the wave?

    A. 2.0×10^{-3} Hz

    B. 60 Hz

    C. 5.0×10^{2} Hz

    D. 1.8×10^{6} Hz

68. An electromagnetic wave propagates through a vacuum. Independent of its wavelength, it will move with constant

    A. Acceleration

    B. Velocity

    C. Induction

    D. Sound

69. A wave generator is used to create a succession of waves. The rate of wave generation is one every 0.33 seconds. The period of these waves is

    A. 2.0 seconds

    B. 1.0 seconds

    C. 0.33 seconds

    D. 3.0 seconds

70. In a fission reactor, heavy water

    A. Cools off neutrons to control temperature

    B. Moderates fission reactions

    C. Initiates the reaction chain

    D. Dissolves control rods

71. Heat transfer by electromagnetic waves is termed

    A. Conduction

    B. Convection

    C. Radiation

    D. Phase Change

72. Solids expand when heated because

    A. Molecular motion causes expansion

    B. $PV = nRT$

    C. Magnetic forces stretch the chemical bonds

    D. All material is effectively fluid

73. Gravitational force at the earth's surface causes

    A. All objects to fall with equal acceleration, ignoring air resistance

    B. Some objects to fall with constant velocity, ignoring air resistance

    C. A kilogram of feathers to float at a given distance above the earth

    D. Aerodynamic objects to accelerate at an increasing rate

74. An office building entry ramp uses the principle of which simple machine?

    A. Lever

    B. Pulley

    C. Wedge

    D. Inclined Plane

75. The velocity of sound is greatest in

    A. Water

    B. Steel

    C. Alcohol

    D. Air

76. All of the following phenomena are considered "refractive effects" except for

    A. The red shift

    B. Total internal reflection

    C. Lens dependent image formation

    D. Snell's Law

77. Static electricity generation occurs by

   A. Telepathy

   B. Friction

   C. Removal of heat

   D. Evaporation

78. The wave phenomenon of polarization applies only to

   A. Longitudinal waves

   B. Transverse waves

   C. Sound

   D. Light

79. A force is given by the vector 5 N x + 3 N y (where x and y are the unit vectors for the x- and y- axes, respectively). This force is applied to move a 10 kg object 5 m, in the x direction. How much work was done?

   A. 250 J

   B. 400 J

   C. 40 J

   D. 25 J

80. A satellite is in a circular orbit above the earth. Which statement is false?

   A. An external force causes the satellite to maintain orbit.

   B. The satellite's inertia causes it to maintain orbit.

   C. The satellite is accelerating toward the earth.

   D. The satellite's velocity and acceleration are not in the same direction.

Answer Key

1.	B	17.	A	33.	A	49.	B	65.	A
2.	B	18.	A	34.	D	50.	B	66.	B
3.	A	19.	B	35.	D	51.	C	67.	A
4.	C	20.	D	36.	B	52.	C	68.	B
5.	C	21.	C	37.	B	53.	A	69.	C
6.	D	22.	C	38.	A	54.	A	70.	B
7.	D	23.	C	39.	A	55.	B	71.	C
8.	A	24.	C	40.	C	56.	C	72.	A
9.	C	25.	C	41.	D	57.	D	73.	A
10.	D	26.	A	42.	A	58.	C	74.	D
11.	D	27.	B	43.	D	59.	B	75.	B
12.	B	28.	B	44.	D	60.	A	76.	A
13.	D	29.	C	45.	D	61.	B	77.	B
14.	A	30.	B	46.	A	62.	D	78.	B
15.	B	31.	C	47.	B	63.	D	79.	D
16.	B	32.	D	48.	D	64.	B	80.	B

Rationales with Sample Questions

1. **A projectile with a mass of 1.0 kg has a muzzle velocity of 1500.0 m/s when it is fired from a cannon with a mass of 500.0 kg. If the cannon slides on a frictionless track, it will recoil with a velocity of ____ m/s.**

    A. 2.4

    B. 3.0

    C. 3.5

    D. 1500

Answer:

B. 3.0

To solve this problem, apply Conservation of Momentum to the cannon-projectile system. The system is initially at rest, with total momentum of 0 kg m/s. Since the cannon slides on a frictionless track, we can assume that the net momentum stays the same for the system. Therefore, the momentum forward (of the projectile) must equal the momentum backward (of the cannon). Thus:

$p_{projectile} = p_{cannon}$
$m_{projectile} \, v_{projectile} = m_{cannon} \, v_{cannon}$
$(1.0 \text{ kg})(1500.0 \text{ m/s}) = (500.0 \text{ kg})(x)$
$x = 3.0 \text{ m/s}$
Only answer (B) matches these calculations.

2. **The weight of an object on the earth's surface is designated _x_. When it is two earth's radii from the surface of the earth, its weight will be**

  A. _x_/4

  B. _x_/9

  C. 4_x_

  D. 16_x_

Answer:

B. _x_/9

  To solve this problem, apply the universal Law of Gravitation to the object and Earth:

  $F_{gravity} = (GM_1M_2)/R^2$
  Because the force of gravity varies with the square of the radius between the objects, the force (or weight) on the object will be decreased by the square of the multiplication factor on the radius. Note that the object on Earth's surface is _already_ at one radius from Earth's center. Thus, when it is two radii from Earth's surface, it is three radii from Earth's center. R^2 is then nine, so the weight is _x_/9.
  Only answer (B) matches these calculations.

3. **If a force of magnitude *F* gives a mass *M* an acceleration *A*, then a force 3*F* would give a mass 3*M* an acceleration**

   A. *A*

   B. 12*A*

   C. *A*/2

   D. 6*A*

Answer:

A. *A*

To solve this problem, apply Newton's Second Law, which is also implied by the first part of the problem:
Force = (Mass)(Acceleration)
$F = MA$
Then apply the same law to the second case, and isolate the unknown:
$3F = 3M\,x$
$x = (3F)/(3M)$
$x = F/M$
$x = A$ (by substituting from our first equation)
Only answer (A) matches these calculations.

4. **A car (mass m_1) is driving at velocity v, when it smashes into an unmoving car (mass m_2), locking bumpers. Both cars move together at the same velocity. The common velocity will be given by**

    A. m_1v/m_2

    B. m_2v/m_1

    C. $m_1v/(m_1 + m_2)$

    D. $(m_1 + m_2)v/m_1$

Answer:

C. $m_1v/(m_1 + m_2)$

In this problem, there is an inelastic collision, so the best method is to assume that momentum is conserved. (Recall that momentum is equal to the product of mass and velocity.)
Therefore, apply Conservation of Momentum to the two-car system:
Momentum at Start = Momentum at End
(Mom. of Car 1) + (Mom. of Car 2) = (Mom. of 2 Cars Coupled)
$m_1v + 0 = (m_1 + m_2)x$
$x = m_1v/(m_1 + m_2)$
Only answer (C) matches these calculations.

Watch out for the other answers, because errors in algebra could lead to a match with incorrect answer (D), and assumption of an elastic collision could lead to a match with incorrect answer (A).

5. **When acceleration is plotted versus time, the area under the graph represents**

    A. Time

    B. Distance

    C. Velocity

    D. Acceleration

Answer:

C. Velocity

The area under a graph will have units equal to the product of the units of the two axes. (To visualize this, picture a graphed rectangle with its area equal to length times width.)
Therefore, multiply units of acceleration by units of time:
$(\text{length}/\text{time}^2)(\text{time})$
This equals length/time, i.e. units of velocity.

6. **An inclined plane is tilted by gradually increasing the angle of elevation θ, until the block will slide down at a constant velocity. The coefficient of friction, μ_k, is given by**

    A. $\cos\theta$

    B. $\sin\theta$

    C. cosecant θ

    D. tangent θ

Answer:

D. tangent θ

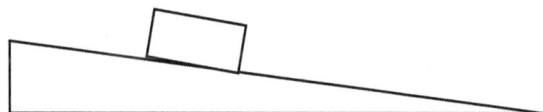

When the block moves, its force upstream (due to friction) must equal its force downstream (due to gravity).

The friction force is given by
$F_f = \mu_k N$
where μ_k is the friction coefficient and N is the normal force.

Using similar triangles, the gravity force is given by
$F_g = mg\sin\theta$
and the normal force is given by
$N = mg\cos\theta$

When the block moves at constant velocity, it must have zero net force, so set equal the force of gravity and the force due to friction:
$F_f = F_g$
$\mu_k\, mg\cos\theta = mg\sin\theta$
$\mu_k = \tan\theta$

Answer (D) is the only appropriate choice in this case.

7. Use the information on heats below to solve this problem. An ice block at 0° Celsius is dropped into 100 g of liquid water at 18° Celsius. When thermal equilibrium is achieved, only liquid water at 0° Celsius is left. What was the mass, in grams, of the original block of ice?

Given: Heat of fusion of ice = 80 cal/g
            Heat of vaporization of ice = 540 cal/g
            Specific Heat of ice = 0.50 cal/g°C
            Specific Heat of water = 1 cal/g°C

   A. 2.0

   B. 5.0

   C. 10.0

   D. 22.5

Answer:

D. 22.5

To solve this problem, apply Conservation of Energy to the ice-water system. Any gain of heat to the melting ice must be balanced by loss of heat in the liquid water. Use the two equations relating temperature, mass, and energy:
$Q = m\,C\,\Delta T$ (for heat loss/gain from change in temperature)
$Q = m\,L$ (for heat loss/gain from phase change)
where Q is heat change; m is mass; C is specific heat; ΔT is change in temperature; L is heat of phase change (in this case, melting, also known as "fusion").

Then
$Q_{ice\ to\ water} = Q_{water\ to\ ice}$
(Note that the ice only melts; it stays at 0° Celsius—otherwise, we would have to include a term for warming the ice as well. Also the information on the heat of vaporization for water is irrelevant to this problem.)
$m\,L = m\,C\,\Delta T$
x (80 cal/g) = 100g 1cal/g°C 18°C
x (80 cal/g) = 1800 cal
x = 22.5 g

Only answer (D) matches this result.

8. **The combination of overtones produced by a musical instrument is known as its**

   A. Timbre

   B. Chromaticity

   C. Resonant Frequency

   D. Flatness

Answer:

A. Timbre

To answer this question, you must know some basic physics vocabulary. "Timbre" is the combination of tones that make a sound unique, beyond its pitch and volume. (For instance, consider the same note played at the same volume, but by different instruments.) Answer (A) is therefore the only appropriate choice. "Resonant Frequency" is relevant to music, because the resonant frequency of a wave will give the dominant sound tone. "Chromaticity" is an analogous word to "timbre," but it describes color tones. "Flatness" is unrelated, and incorrect.

9. **A long copper bar has a temperature of 60°C at one end and 0°C at the other. The bar reaches thermal equilibrium (barring outside influences) by the process of heat**

   A. Fusion

   B. Convection

   C. Conduction

   D. Microwaving

Answer:

C. Conduction

To answer this question, recall the different methods of heat transfer. (Note that since the bar is warm at one end and cold at the other, heat must transfer through the bar from warm to cold, until temperature is equalized.) "Convection" is the heat transfer via fluid currents. "Conduction" is the heat transfer via connected solid material. "Fusion" and "Microwaving" are not methods of heat transfer. Therefore the only appropriate answer is (C).

10. **The First Law of Thermodynamics takes the form dU = dW when the conditions are**

    A. Isobaric

    B. Isochloremic

    C. Isothermal

    D. Adiabatic

Answer:

D. Adiabatic

To answer this question, recall the First Law of Thermodynamics:
Change in Internal Energy = Work Done + Heat Added
dU = dW + dQ

Thus in the form we are given, dQ has been set to zero, i.e. there is no heat added.
"Adiabatic" refers to a case where there is no heat exchange with surroundings, so answer (D) is the appropriate choice. "Isobaric" means at a constant pressure, "Isothermal" means at a constant temperature, and "Isochloremic" is an imaginary word, as far as I can tell.

It might be tempting to choose "Isothermal," thinking that no heat added would require the same temperature. However, work and internal energy changes can change temperature within the system analyzed, even when no heat is exchanged with the surroundings.

11. Given a vase full of water, with holes punched at various heights. The water squirts out of the holes, achieving different distances before hitting the ground. Which of the following accurately describes the situation?

    A. Water from higher holes goes farther, due to Pascal's Principle.

    B. Water from higher holes goes farther, due to Bernoulli's Principle.

    C. Water from lower holes goes farther, due to Pascal's Principle.

    D. Water from lower holes goes farther, due to Bernoulli's Principle.

Answer:

D. Water from lower holes goes farther, due to Bernoulli's Principle.

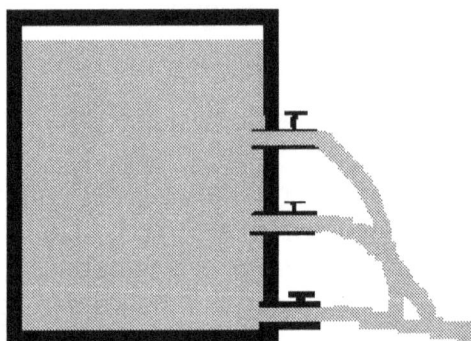

To answer this question, consider the pressure on the water in the vase. The deeper the water, the higher the pressure. Thus, when a hole is punched, the water stream will achieve higher velocity as it equalizes to atmospheric pressure. The lower streams will therefore travel farther before hitting the ground. This eliminates answers (A) and (B). Then recall that Pascal's Principle provides for immediate pressure changes throughout a fluid, while Bernoulli's Principle translates pressure, velocity, and height energy into each other. In this case, the pressure energy is being transformed into velocity energy, and Bernoulli's Principle applies. Therefore, the only appropriate answer is (D).

12. A stationary sound source produces a wave of frequency F. An observer at position A is moving toward the horn, while an observer at position B is moving away from the horn. Which of the following is true?

    A. $F_A < F < F_B$

    B. $F_B < F < F_A$

    C. $F < F_A < F_B$

    D. $F_B < F_A < F$

Answer:

B. $F_B < F < F_A$

To answer this question, recall the Doppler Effect. As a moving observer approaches a sound source, s/he intercepts wave fronts sooner than if s/he were standing still. Therefore, the wave fronts seem to be coming more frequently. Similarly, as an observer moves away from a sound source, the wave fronts take longer to reach him/her. Therefore, the wave fronts seem to be coming less frequently. Because of this effect, the frequency at B will seem lower than the original frequency, and the frequency at A will seem higher than the original frequency. The only answer consistent with this is (B). Note also, that even if you weren't sure of which frequency should be greater/smaller, you could still reason that A and B should have opposite effects, and be able to eliminate answer choices (C) and (D).

13. The electric force in Newtons, on two small objects (each charged to −10 microCoulombs and separated by 2 meters) is

    A. 1.0

    B. 9.81

    C. 31.0

    D. 0.225

Answer:

D. 0.225

To answer this question, use Coulomb's Law, which gives the electric force between two charged particles:
$F = k Q_1 Q_2 / r^2$
Then our unknown is F, and our knowns are:
$k = 9.0 \times 10^9$ Nm^2/C^2
$Q_1 = Q_2 = -10 \times 10^{-6}$ C
$r = 2$ m

Therefore
$F = (9.0 \times 10^9)(-10 \times 10^{-6})(-10 \times 10^{-6})/(2^2)$ N
$F = 0.225$ N

This is compatible only with answer (D).

14. A 10 ohm resistor and a 50 ohm resistor are connected in parallel. If the current in the 10 ohm resistor is 5 amperes, the current (in amperes) running through the 50 ohm resistor is

    A. 1

    B. 50

    C. 25

    D. 60

Answer:

A. 1

To answer this question, use Ohm's Law, which relates voltage to current and resistance:
$V = IR$
where V is voltage; I is current; R is resistance.

We also use the fact that in a parallel circuit, the voltage is the same across the branches.

Because we are given that in one branch, the current is 5 amperes and the resistance is 10 ohms, we deduce that the voltage in this circuit is their product, 50 volts (from $V = IR$).

We then use $V = IR$ again, this time to find I in the second branch. Because V is 50 volts, and R is 50 ohm, we calculate that I has to be 1 ampere.

This is consistent only with answer (A).

15. Fahrenheit and Celsius thermometers have the same temperature reading at

    A. 100 degrees

    B. - 40 degrees

    C. Absolute Zero

    D. 40 degrees

Answer:

B. - 40 degrees

To answer this question, use the relationship between Fahrenheit and Celsius temperature scales:
$$F = 9/5 \ C + 32$$

Then, in a case where both °F and °C are equal, F = C, so
$$C = 9/5 \ C + 32$$
$$- 4/5 \ C = 32$$
$$C = - 40$$

Only answer (B) is consistent with this result.

16. Which of the following apparatus can be used to measure the wavelength of a sound produced by a tuning fork?

        A. A glass cylinder, some water, and iron filings

        B. A glass cylinder, a meter stick, and some water

        C. A metronome and some ice water

        D. A comb and some tissue

Answer:

B. A glass cylinder, a meter stick, and some water

To answer this question, recall that a sound will be amplified if it is reflected back to cause positive interference. This is the principle behind musical instruments that use vibrating columns of air to amplify sound (e.g. a pipe organ).

Therefore, presumably a person could put varying amounts of water in the cylinder, and hold the vibrating tuning fork above the cylinder in each case. If the tuning fork sound is amplified when put at the top of the column, then the length of the air space would be an integral multiple of the sound's wavelength. This experiment is consistent with answer (B). Although the experiment would be tedious, none of the other options for materials suggest a better alternative.

17. When the current flowing through a fixed resistance is doubled, the amount of heat generated is

   A. Quadrupled

   B. Doubled

   C. Multiplied by pi

   D. Halved

Answer:

A. Quadrupled

   To answer this question, recall that heat generated will occur because of the power of the circuit (power is energy per time). For a circuit with a fixed resistance:
   $P = I\,V$
   where P is power; I is current; V is voltage. Then use Ohm's Law:
   $V = I\,R$
   where V is voltage; I is current; R is resistance, and substitute:
   $P = I^2\,R$
   and so the doubling of the current I will lead to a quadrupling of the power, and therefore the a quadrupling of the heat.

   This is consistent only with answer (A). If you weren't sure of the equations, you could still deduce that with more current, there would be more heat generated, and therefore eliminate answer choice (D) in any case.

18. The current induced in a coil is defined by which of the following laws?

    A. Lenz's Law

    B. Burke's Law

    C. The Law of Spontaneous Combustion

    D. Snell's Law

Answer:

A. Lenz's Law

Lenz's Law states that an induced electromagnetic force always gives rise to a current whose magnetic field opposes the original flux change. There is no relevant "Snell's Law," "Burke's Law," or "Law of Spontaneous Combustion" in electromagnetism. (In fact, only Snell's Law is a real law of these three, and it refers to refracted light.) Therefore, the only appropriate answer is (A).

19. If an object is 20 cm from a convex lens whose focal length is 10 cm, the image is:

A. Virtual and upright

B. Real and inverted

C. Larger than the object

D. Smaller than the object

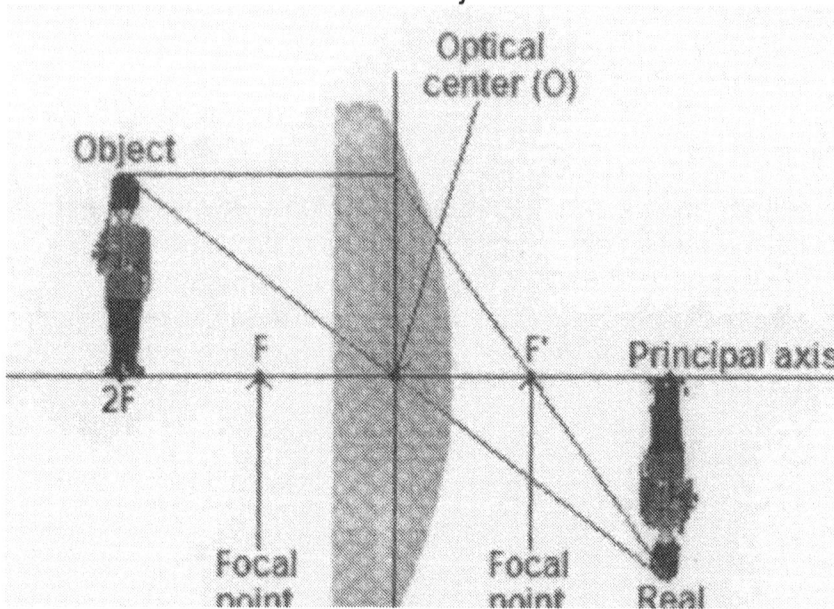

Answer:

B. Real and inverted

To solve this problem, draw a lens diagram with the lens, focal length, and image size.

The ray from the top of the object straight to the lens is focused through the far focus point; the ray from the top of the object through the near focus goes straight through the lens; the ray from the top of the object through the center of the lens continues. These three meet to form the "top" of the image, which is therefore real and inverted. This is consistent only with answer (B).

20. A cooking thermometer in an oven works because the metals it is composed of have different

   A. Melting points

   B. Heat convection

   C. Magnetic fields

   D. Coefficients of expansion

Answer:

D. Coefficients of expansion

A thermometer of the type that can withstand oven temperatures works by having more than one metal strip. These strips expand at different rates with temperature increases, causing the dial to register the new temperature. This is consistent only with answer (D). If you did not know how an oven thermometer works, you could still omit the incorrect answers: It is unlikely that the metals in a thermometer would melt in the oven to display the temperature; the magnetic fields would not be useful information in this context; heat convection applies in fluids, not solids.

21. In an experiment where a brass cylinder is transferred from boiling water into a beaker of cold water with a thermometer in it, we are measuring

    A. Fluid viscosity

    B. Heat of fission

    C. Specific heat

    D. Nonspecific heat

Answer:

C. Specific heat

In this question, we consider an experiment to measure temperature change of water (with the thermometer) as the cylinder cools and the water warms. This information can be used to calculate heat changes, and therefore specific heat. Therefore, (C) is the correct answer. Even if you were unable to deduce that specific heat is being measured, you could eliminate the other answer choices: viscosity cannot be measured with a thermometer; fission takes place at much higher temperatures than this experiment, and under quite different conditions; there is no such thing as "nonspecific heat".

22. A temperature change of 40 degrees Celsius is equal to a change in Fahrenheit degrees of

    A. 40

    B. 20

    C. 72

    D. 112

Answer:

C. 72

To answer this question, recall the equation for Celsius and Fahrenheit:
$°F = 9/5 \ °C + 32$

Therefore, whatever temperature difference occurs in °C, it is multiplied by a factor of 9/5 to get the new °F measurement:
$\text{new}°F = 9/5(\text{old}°C + 40) + 32$
(whereas $\text{old}°F = 9/5(\text{old}°C) + 32$)
Therefore the difference between the old and new temperatures in Fahrenheit is 9/5 of 40, or 72 degrees. This is consistent only with answer (C).

23. The number of calories required to raise the temperature of 40 grams of water at 30°C to steam at 100°C is

     A. 7500

     B. 23,000

     C. 24,400

     D. 30,500

Answer:

C. 24,400

To answer this question, apply the equations for heat transfer due to temperature and phase changes:

$Q = mC\Delta T + mL$

where Q is heat; m is mass; C is specific heat; ΔT is temperature change; L is heat of phase change.

In this problem, we are trying to find Q, and we are given:
m = 40 g
C = 1 cal/g°C for water (this should be memorized)
ΔT = 70 °C
L = 540 cal/g for liquid to gas change in water (this should be memorized)

thus Q = (40 g)(1 cal/g°C)(70 °C) + (40 g)(540 cal/g)
Q = 24,400 cal
This is consistent only with answer (C).

24. The boiling point of water on the Kelvin scale is closest to

   A. 112 K

   B. 212 K

   C. 373 K

   D. 473 K

Answer:

C. 373 K

To answer this question, recall that Kelvin temperatures are equal to Celsius temperatures plus 273.15. Since water boils at 100°C under standard conditions, it will boil at 373.15 K. This is consistent only with answer (C).

25. The kinetic energy of an object is _____ proportional to its
_____.

   A. Inversely...inertia

   B. Inversely...velocity

   C. Directly...mass

   D. Directly...time

Answer:

C. Directly...mass

To answer this question, recall that kinetic energy is equal to one-half of the product of an object's mass and the square of its velocity:
$KE = \frac{1}{2} m v^2$

Therefore, kinetic energy is directly proportional to mass, and the answer is (C). Note that although kinetic energy is associated with both velocity and momentum (a measure of inertia), it is not *inversely* proportional to either one.

26. A hollow conducting sphere of radius R is charged with a total charge Q. What is the magnitude of the electric field at a distance r (given r<R) from the center of the sphere? (k is the electrostatic constant)

    A. 0

    B. $k\,Q/R^2$

    C. $k\,Q/(R^2 - r^2)$

    D. $k\,Q/(R - r)^2$

Answer:

A. 0

You may be tempted to use the equation for the electric field:
$E = F/Q$ (E = electric field; F = electric force; Q = charge)

and the Coulomb's Law expression for electric force:
$F = k\,Q_1 Q_2/r^2$ (k = constant; Q_1 and Q_2 = charges; r = distance apart),

which usually would give
$E = k\,Q/(R\text{-}r)^2$ in a similar context.

However, this question addresses a special case, i.e. a hollow conductor. Inside a hollow conductor, no electric field exists, because if there were an electric field inside, the conductor's free electrons would be forced to move (by the electric force) until the electric field became zero. Therefore, the only correct answer is (A).

27. A quantum of light energy is called a

    A. Dalton

    B. Photon

    C. Curie

    D. Heat Packet

Answer:

B. Photon

The smallest "packet" (quantum) of light energy is a photon. "Heat Packet" does not have any relevant meaning, and while "Dalton" and "Curie" have other meanings, they are not connected to light. Therefore, only (B) is a correct answer.

28. The following statements about sound waves are true *except*

    A. Sound travels faster in liquids than in gases.

    B. Sound waves travel through a vacuum.

    C. Sound travels faster through solids than liquids.

    D. Ultrasound can be reflected by the human body.

Answer:

B. Sound waves travel through a vacuum.

Sound waves require a medium to travel. The sound wave agitates the material, and this occurs fastest in solids, then liquids, then gases. Ultrasound waves are reflected by parts of the body, and this is useful in medical imaging. Therefore, the only correct answer is (B).

29. The greatest number of 100 watt lamps that can be connected in parallel with a 120 volt system without blowing a 5 amp fuse is

    A. 24

    B. 12

    C. 6

    D. 1

Answer:

C. 6

To solve fuse problems, you must add together all the drawn current in the parallel branches, and make sure that it is less than the fuse's amp measure. Because we know that electrical power is equal to the product of current and voltage, we can deduce that:
$I = P/V$ (I = current (amperes); P = power (watts); V = voltage (volts))

Therefore, for each lamp, the current is 100/120 amperes, or 5/6 ampere. The highest possible number of lamps is thus six, because six lamps at 5/6 ampere each adds to 5 amperes; more will blow the fuse.

This is consistent only with answer (C).

30. A monochromatic ray of light passes from air to a thick slab of glass (n = 1.41) at an angle of 45° from the normal. At what angle does it leave the air/glass interface?

    A. 45°

    B. 30°

    C. 15°

    D. 55°

Answer:

B. 30°

To solve this problem use Snell's Law:
$n_1 \sin\theta_1 = n_2 \sin\theta_2$ (where n_1 and n_2 are the indexes of refraction and θ_1 and θ_2 are the angles of incidence and refraction).

Then, since the index of refraction for air is 1.0, we deduce:
$1 \sin 45° = 1.41 \sin x$
$x = \sin^{-1} ((1/1.41) \sin 45°)$
$x = 30°$

This is consistent only with answer (B). Also, note that you could eliminate answers (A) and (D) in any case, because the refracted light will have to bend at a smaller angle when entering glass.

31. The magnitude of a force is

  A. Directly proportional to mass and inversely to acceleration

  B. Inversely proportional to mass and directly to acceleration

  C. Directly proportional to both mass and acceleration

  D. Inversely proportional to both mass and acceleration

Answer:

C. Directly proportional to both mass and acceleration

To solve this problem, recall Newton's 2nd Law, i.e. net force is equal to mass times acceleration. Therefore, the only possible answer is (C).

32. A semi-conductor allows current to flow

  A. Never

  B. Always

  C. As long as it stays below a maximum temperature

  D. When a minimum voltage is applied

Answer:

D. When a minimum voltage is applied

To answer this question, recall that semiconductors do not conduct as well as conductors (eliminating answer (B)), but they conduct better than insulators (eliminating answer (A)). Semiconductors can conduct better when the temperature is higher (eliminating answer (C)), and their electrons move most readily under a potential difference. Thus the answer can only be (D).

33. One reason to use salt for melting ice on roads in the winter is that

      A. Salt lowers the freezing point of water.

      B. Salt causes a foaming action, which increases traction.

      C. Salt is more readily available than sugar.

      D. Salt increases the conductivity of water.

Answer:

A. Salt lowers the freezing point of water.

In answering this question, you may recall that salt is used for road traction because it has large particles to increase traction (analogous to sand). This is true, but salt also has the potential to lower the freezing point of water, thus melting the ice with which it has contact. This is consistent with answer (A). Answer (B) is untrue, and the other two choices, while usually true, are irrelevant in this case.

34. Automobile mirrors that have a sign, "objects are closer than they appear" say so because

      A. The real image of an obstacle, through a converging lens, appears farther away than the object.

      B. The real or virtual image of an obstacle, through a converging mirror, appears farther away than the object.

      C. The real image of an obstacle, through a diverging lens, appears farther away than the object.

      D. The virtual image of an obstacle, through a diverging mirror, appears farther away than the object.

Answer:

D. The virtual image of an obstacle, through a diverging mirror, appears farther away than the object.

To answer this question, first eliminate answer choices (A) and (C), because we have a mirror, not a lens. Then draw ray diagrams for diverging (convex) and converging (concave) mirrors, and note that because the focal point of a diverging mirror is behind the surface, the image is smaller than the object. This creates the illusion that the object is farther away, and therefore (D) is the correct answer.

35. **A gas maintained at a constant pressure has a specific heat which is greater than its specific heat when maintained at constant volume, because**

    A. The Coefficient of Expansion changes.

    B. The gas enlarges as a whole.

    C. Brownian motion causes random activity.

    D. Work is done to expand the gas.

Answer:

D. Work is done to expand the gas.

To answer this question, recall that the specific heat is a measure of how much energy it takes to raise the temperature of a given mass of gas. Thus, you can reason that when a gas is maintained at constant pressure, some energy is used to expand the volume of the gas, and less is left for temperature changes. In fact, this is the case, and (D) is the correct answer. If you were not able to figure that out, you could still eliminate the other answers, because they are not strictly relevant to a change in specific heat.

36. **Consider the shear modulus of water, and that of mercury. Which of the following is true?**

    A. Mercury's shear modulus indicates that it is the only choice of fluid for a thermometer.

    B. The shear modulus of each of these is zero.

    C. The shear modulus of mercury is higher than that of water.

    D. The shear modulus of water is higher than that of mercury.

Answer:

B. The shear modulus of each of these is zero.

To answer this question, recall that shear modulus is meaningful only for solids, and that liquids, instead, have a bulk modulus. The only reasonable answer is therefore (B).

37. A skateboarder accelerates down a ramp, with constant acceleration of two meters per second squared, from rest. The distance in meters, covered after four seconds, is

    A. 10

    B. 16

    C. 23

    D. 37

Answer:

B. 16

To answer this question, recall the equation relating constant acceleration to distance and time:
$x = \frac{1}{2} a t^2 + v_0 t + x_0$ where x is position; a is acceleration; t is time; v_0 and x_0 are initial velocity and position (both zero in this case)

thus, to solve for x:
$x = \frac{1}{2} (2 \text{ m/s}^2) (4^2 s^2) + 0 + 0$
$x = 16 \text{ m}$

This is consistent only with answer (B).

38. Which of the following units is not used to measure torque?

    A. slug ft

    B. lb ft

    C. N m

    D. dyne cm

Answer:

A. slug ft

To answer this question, recall that torque is always calculated by multiplying units of force by units of distance. Therefore, answer (A), which is the product of units of mass and units of distance, must be the choice of incorrect units. Indeed, the other three answers all could measure torque, since they are of the correct form. It is a good idea to review "English Units" before the teacher test, because they are occasionally used in problems.

39. In a nuclear pile, the control rods are composed of

    A. Boron

    B. Einsteinium

    C. Isoptocarpine

    D. Phlogiston

Answer:

A. Boron

Nuclear plants use control rods made of boron or cadmium, to absorb neutrons and maintain "critical" conditions in the reactor. However, if you did not know that, you could still eliminate choice (D), because "phlogiston" is the word for the imaginary element in an ancient structure of earth-air-water-fire.

40. All of the following use semi-conductor technology, except a(n):

    A. Transistor

    B. Diode

    C. Capacitor

    D. Operational Amplifier

Answer:

C. Capacitor

Semi-conductor technology is used in transistors and operational amplifiers, and diodes are the basic unit of semi-conductors. Therefore the only possible answer is (C), and indeed a capacitor does not require semi-conductor technology.

41. Ten grams of a sample of a radioactive material (half-life = 12 days) were stored for 48 days and re-weighed. The new mass of material was

    A. 1.25 g

    B. 2.5 g

    C. 0.83 g

    D. 0.625 g

Answer:

D. 0.625 g

To answer this question, note that 48 days is four half-lives for the material. Thus, the sample will degrade by half four times. At first, there are ten grams, then (after the first half-life) 5 g, then 2.5 g, then 1.25 g, and after the fourth half-life, there remains 0.625 g. You could also do the problem mathematically, by multiplying ten times $(\frac{1}{2})^4$, i.e. $\frac{1}{2}$ for each half-life elapsed.

42. When a radioactive material emits an alpha particle only, its atomic number will

   A. Decrease

   B. Increase

   C. Remain unchanged

   D. Change randomly

Answer:

A. Decrease

   To answer this question, recall that in alpha decay, a nucleus emits the equivalent of a Helium atom. This includes two protons, so the original material changes its atomic number by a decrease of two.

43. The sun's energy is produced primarily by

   A. Fission

   B. Explosion

   C. Combustion

   D. Fusion

Answer:

D. Fusion

   To answer this question, recall that in stars (such as the sun), fusion is the main energy-producing occurrence. Fission, explosion, and combustion all release energy in other contexts, but they are not the right answers here.

44. A crew is on-board a spaceship, traveling at 60% of the speed of light with respect to the earth. The crew measures the length of their ship to be 240 meters. When a ground-based crew measures the apparent length of the ship, it equals

    A. 400 m

    B. 300 m

    C. 240 m

    D. 192 m

Answer:

D. 192 m

To answer this question, recall that a moving object's size seems contracted to the stationary observer, according to the equation:
Length Observed = (Actual Length) $(1 - v^2/c^2)^{\frac{1}{2}}$
where v is speed of motion, and c is speed of light.

Therefore, in this case,
Length Observed = (240 m) $(1 - 0.6^2)^{\frac{1}{2}}$
Length Observed = (240 m) (0.8) = 192 m

This is consistent only with answer (D). If you were unsure of the equation, you could still reason that because of length contraction (the flip side of time dilation), you must choose an answer with a smaller length, and only (D) fits that description. Note that only the dimension in the direction of travel is contracted. (The length in this case.)

45. Which of the following pairs of elements are not found to fuse in the centers of stars?

    A. Oxygen and Helium

    B. Carbon and Hydrogen

    C. Beryllium and Helium

    D. Cobalt and Hydrogen

Answer:

D. Cobalt and Hydrogen

To answer this question, recall that fusion is possible only when the final product has more binding energy than the reactants. Because binding energy peaks near a mass number of around 56, corresponding to Iron, any heavier elements would be unlikely to fuse in a typical star. (In very massive stars, there may be enough energy to fuse heavier elements.) Of all the listed elements, only Cobalt is heavier than iron, so answer (D) is correct.

46. A calorie is the amount of heat energy that will

    A. Raise the temperature of one gram of water from $14.5°$ C to $15.5°$ C.

    B. Lower the temperature of one gram of water from $16.5°$ C to $15.5°$ C

    C. Raise the temperature of one gram of water from $32°$ F to $33°$ F

    D. Cause water to boil at two atmospheres of pressure.

Answer:

A. Raise the temperature of one gram of water from $14.5°$ C to $15.5°$ C.

The definition of a calorie is, "the amount of energy to raise one gram of water by one degree Celsius," and so answer (A) is correct. Do not get confused by the fact that $14.5°$ C seems like a random number. Also, note that answer (C) tries to confuse you with degrees Fahrenheit, which are irrelevant to this problem.

47. Bohr's theory of the atom was the first to quantize

    A. Work

    B. Angular Momentum

    C. Torque

    D. Duality

Answer:

B. Angular Momentum

    Bohr was the first to quantize the angular momentum of electrons, as he combined Rutherford's planet-style model with his knowledge of emerging quantum theory. Recall that he derived a "quantum condition" for the single electron, requiring electrons to exist at specific energy levels.

48. A uniform pole weighing 100 grams, that is one meter in length, is supported by a pivot at 40 centimeters from the left end. In order to maintain static position, a 200 gram mass must be placed _____ \centimeters from the left end.

    A. 10

    B. 45

    C. 35

    D. 50

Answer:

D. 50

    In answering this question, do not be tricked into calculating the position of the mass to create balance on the pole's pivot. (This calculation, with equal torques, would lead to incorrect answer (B).) A careful read of the question reveals that we want to "maintain static position" i.e. keep the pole from moving. Because it is already tilted toward the right side, that side (or anywhere to the right of the 45 cm pivot balance answer) is the correct place to put the additional weight without causing the pole to move. Thus, only answer (D) can be correct.

49. **A classroom demonstration shows a needle floating in a tray of water. This demonstrates the property of**

   A. Specific Heat

   B. Surface Tension

   C. Oil-Water Interference

   D. Archimedes' Principle

Answer:

B. Surface Tension

To answer this question, note that the only information given is that the needle (a small object) floats on the water. This occurs because although the needle is denser than the water, the surface tension of the water causes sufficient resistance to support the small needle. Thus the answer can only be (B). Answer (A) is unrelated to objects floating, and while answers (C) and (D) could be related to water experiments, they are not correct in this case. There is no oil in the experiment, and Archimedes' Principle allows the equivalence of displaced volumes, which is not relevant here.

50. **Two neutral isotopes of a chemical element have the same numbers of**

   A. Electrons and Neutrons

   B. Electrons and Protons

   C. Protons and Neutrons

   D. Electrons, Neutrons, and Protons

Answer:

B. Electrons and Protons

To answer this question, recall that isotopes vary in their number of neutrons. (This fact alone eliminates answers (A), (C), and (D).) If you did not recall that fact, note that we are given that the two samples are of the same element, constraining the number of protons to be the same in each case. Then, use the fact that the samples are neutral, so the number of electrons must exactly balance the number of protons in each case. The only correct answer is thus (B).

51. A mass is moving at constant speed in a circular path. Choose the true statement below:

    A. Two forces in equilibrium are acting on the mass.

    B. No forces are acting on the mass.

    C. One centripetal force is acting on the mass.

    D. One force tangent to the circle is acting on the mass.

Answer:

C. One centripetal force is acting on the mass.

To answer this question, recall that by Newton's 2nd Law, $F = ma$. In other words, force is mass times acceleration. Furthermore, acceleration is any change in the velocity vector—whether in size or direction. In circular motion, the direction of velocity is constantly changing. Therefore, there must be an unbalanced force on the mass to cause that acceleration. This eliminates answers (A) and (B) as possibilities. Recall then that the mass would ordinarily continue traveling tangent to the circle (by Newton's 1st Law). Therefore, the force must be to cause the turn, i.e. a centripetal force. Thus, the answer can only be (C).

52. A light bulb is connected in series with a rotating coil within a magnetic field. The brightness of the light may be increased by any of the following except:

    A. Rotating the coil more rapidly.

    B. Using more loops in the coil.

    C. Using a different color wire for the coil.

    D. Using a stronger magnetic field.

Answer:

C. Using a different color wire for the coil.

To answer this question, recall that the rotating coil in a magnetic field generates electric current, by Faraday's Law. Faraday's Law states that the amount of emf generated is proportional to the rate of change of magnetic flux through the loop. This increases if the coil is rotated more rapidly (A), if there are more loops (B), or if the magnetic field is stronger (D). Thus, the only answer to this question is (C).

53. The use of two circuits next to each other, with a change in current in the primary circuit, demonstrates

    A. Mutual current induction

    B. Dielectric constancy

    C. Harmonic resonance

    D. Resistance variation

Answer:

A. Mutual current induction

To answer this question, recall that changing current induces a change in magnetic flux, which in turn causes a change in current to oppose that change (Lenz's and Faraday's Laws). Thus, (A) is correct. If you did not remember that, note that harmonic resonance is irrelevant here (eliminating (C)), and there is no change in resistance in the circuits (eliminating (D)).

54. A brick and hammer fall from a ledge at the same time. They would be expected to

    A. Reach the ground at the same time

    B. Accelerate at different rates due to difference in weight

    C. Accelerate at different rates due to difference in potential energy

    D. Accelerate at different rates due to difference in kinetic energy

Answer:

A. Reach the ground at the same time

This is a classic question about falling in a gravitational field. All objects are acted upon equally by gravity, so they should reach the ground at the same time. (In real life, air resistance can make a difference, but not at small heights for similarly shaped objects.) In any case, weight, potential energy, and kinetic energy do not affect gravitational acceleration. Thus, the only possible answer is (A).

55. The potential difference across a five Ohm resistor is five Volts. The power used by the resistor, in Watts, is

    A. 1

    B. 5

    C. 10

    D. 20

Answer:

B. 5

To answer this question, recall the two relevant equations for potential difference and electric power:
$V = IR$ (where V is voltage; I is current; R is resistance)
$P = IV = I^2R$ (where P is power; I is current; R is resistance)

Thus, first calculate the current from the first equation:
$I = V/R = 1$ Ampere

And then use the second equation:
$P = I^2R = 5$ Watts

This is consistent only with answer (B).

56. An object two meters tall is speeding toward a plane mirror at 10 m/s. What happens to the image as it nears the surface of the mirror?

    A. It becomes inverted.

    B. The Doppler Effect must be considered.

    C. It remains two meters tall.

    D. It changes from a real image to a virtual image.

Answer:

C. It remains two meters tall.

Note that the mirror is a plane mirror, so the image is always a virtual image of the same size as the object. If the mirror were concave, then the image would be inverted until the object came within the focal distance of the mirror. The Doppler Effect is not relevant here. Thus, the only possible answer is (C).

57. The highest energy is associated with

    A. UV radiation

    B. Yellow light

    C. Infrared radiation

    D. Gamma radiation

Answer:

D. Gamma radiation

To answer this question, recall the electromagnetic spectrum. The highest energy (and therefore frequency) rays are those with the lowest wavelength, i.e. gamma rays. (In order of frequency from lowest to highest are: radio, microwave, infrared, red through violet visible light, ultraviolet, X-rays, gamma rays.) Thus, the only possible answer is (D). Note that even if you did not remember the spectrum, you could deduce that gamma radiation is considered dangerous and thus might have the highest energy.

58. The constant of proportionality between the energy and the frequency of electromagnetic radiation is known as the

    A. Rydberg constant

    B. Energy constant

    C. Planck constant

    D. Einstein constant

Answer:

C. Planck constant

Planck estimated his constant to determine the ratio between energy and frequency of radiation. The Rydberg constant is used to find the wavelengths of the visible lines on the hydrogen spectrum.
The other options are not relevant options, and may not actually have physical meaning. Therefore, the only possible answer is (C).

59. A simple pendulum with a period of one second has its mass doubled. If the length of the string is quadrupled, the new period will be

    A. 1 second

    B. 2 seconds

    C. 3 seconds

    D. 5 seconds

Answer:

B. 2 seconds

To answer this question, recall that the period of a pendulum is given by:

$T = 2\pi (L/g)^{1/2}$ where T is period; L is length; g is gravitational acceleration (This is derived from balancing the forces and making small-angle approximation for small angles.)
Note that this equation is independent of mass, so that change is irrelevant. Since the length is quadrupled, and all other quantities on the right side of the equation are constant, the new period will be increased by a factor of two (the square root of four).
This is consistent only with answer (B).

60. A vibrating string's frequency is _____ proportional to the _____.

    A. Directly; Square root of the tension

    B. Inversely; Length of the string

    C. Inversely; Squared length of the string

    D. Inversely; Force of the plectrum

Answer:

A. Directly; Square root of the tension

To answer this question, recall that
$f = (n\ v) / (2\ L)$ where f is frequency; v is velocity; L is length

and

$v = (F_{tension} / (m / L))^{1/2}$ where $F_{tension}$ is tension; m is mass; others as above

so

$f = (n / 2\ L) ((F_{tension} / (m / L))^{1/2}\ )$

indicating that frequency is directly proportional to the square root of the tension force. This is consistent only with answer (A). Note that in the final frequency equation, there is an inverse relationship with the square root of the length (after canceling like terms). This is not one of the options, however.

61. When an electron is "orbiting" the nucleus in an atom, it is said to possess an intrinsic spin (spin angular momentum). How many values can this spin have in any given electron?

> A. 1
>
> B. 2
>
> C. 3
>
> D. 8

Answer:

B. 2

> To answer this question, recall that electrons fill orbitals in pairs, and the two electrons in any pair have opposite spin from one another. Thus, (B) is correct. Note that answer (D) is trying to mislead you into thinking of the number of valence electrons in an atom.

62. Electrons are

> A. More massive than neutrons
>
> B. Positively charged
>
> C. Neutrally charged
>
> D. Negatively charged

Answer:

D. Negatively charged

> Electrons are negatively charged particles that have a tiny mass compared to protons and neutrons. Thus, answer (D) is the only correct alternative.

63. Rainbows are created by

    A. Reflection, dispersion, and recombination

    B. Reflection, resistance, and expansion

    C. Reflection, compression, and specific heat

    D. Reflection, refraction, and dispersion

Answer:

D. Reflection, refraction, and dispersion

To answer this question, recall that rainbows are formed by light that goes through water droplets and is dispersed into its colors. This is consistent with both answers (A) and (D). Then note that refraction is important in bending the differently colored light waves, while recombination is not a relevant concept here. Therefore, the answer is (D).

64. In order to switch between two different reference frames in special relativity, we use the _____ transformation.

    A. Galilean

    B. Lorentz

    C. Euclidean

    D. Laplace

Answer:

B. Lorentz

The Lorentz transformation is the set of equations to scale length and time between inertial reference frames in special relativity, when velocities are close to the speed of light.

The Galilean transformation is a parallel set of equations, used for 'classical' situations when velocities are much slower than the speed of light. Euclidean geometry is useful in physics, but not relevant here. Laplace transforms are a method of solving differential equations by using exponential functions. The correct answer is therefore (B).

65. **A baseball is thrown with an initial velocity of 30 m/s at an angle of 45°. Neglecting air resistance, how far away will the ball land?**

    A. 92 m

    B. 78 m

    C. 65 m

    D. 46 m

Answer:

A. 92 m

To answer this question, recall the equations for projectile motion:
$y = \frac{1}{2} a t^2 + v_{0y} t + y_0$
$x = v_{0x} t + x_0$
where x and y are horizontal and vertical position, respectively; t is time; a is acceleration due to gravity; v_{0x} and v_{0y} are initial horizontal and vertical velocity, respectively; x_0 and y_0 are initial horizontal and vertical position, respectively.
For our case:
x_0 and y_0 can be set to zero
both v_{0x} and v_{0y} are (using trigonometry) = $(\sqrt{2} / 2)$ 30 m/s
$a = -9.81$ m/s^2

We then use the vertical motion equation to find the time aloft (setting y equal to zero to find the solution for t):
$0 = \frac{1}{2} (-9.81 \text{ m/s}^2) t^2 + (\sqrt{2} / 2)$ 30 m/s t
Then solving, we find:
t = 0 s (initial set-up) or t = 4.324 s (time to go up and down)

Using t = 4.324 s in the horizontal motion equation, we find:
$x = ((\sqrt{2} / 2)$ 30 m/s) (4.324 s)
x = 91.71 m

This is consistent only with answer (A).

66. If one sound is ten decibels louder than another, the ratio of the intensity of the first to the second is

    A. 20:1

    B. 10:1

    C. 1:1

    D. 1:10

Answer:

B. 10:1

To answer this question, recall that a decibel is defined as ten times the log of the ratio of sound intensities:
(decibel measure) = $10 \log (I / I_0)$ where I_0 is a reference intensity.

Therefore, in our case,
(decibels of first sound) = (decibels of second sound) + 10
$10 \log (I_1 / I_0) = 10 \log (I_2 / I_0) + 10$
$10 \log I_1 - 10 \log I_0 = 10 \log I_2 - 10 \log I_0 + 10$
$10 \log I_1 - 10 \log I_2 = 10$
$\log (I_1 / I_2) = 1$
$I_1 / I_2 = 10$

This is consistent only with answer (B).
(Be careful not to get the two intensities confused with each other.)

67. A wave has speed 60 m/s and wavelength 30,000 m. What is the frequency of the wave?

   A. 2.0×10^{-3} Hz

   B. 60 Hz

   C. 5.0×10^{2} Hz

   D. 1.8×10^{6} Hz

Answer:

A. 2.0×10^{-3} Hz

   To answer this question, recall that wave speed is equal to the product of wavelength and frequency. Thus:
   60 m/s = (30,000 m) (frequency)
   frequency = 2.0×10^{-3} Hz

   This is consistent only with answer (A).

68. An electromagnetic wave propagates through a vacuum. Independent of its wavelength, it will move with constant

   A. Acceleration

   B. Velocity

   C. Induction

   D. Sound

Answer:

B. Velocity

   Electromagnetic waves are considered always to travel at the speed of light, so answer (B) is correct. Answers (C) and (D) can be eliminated in any case, because induction is not relevant here, and sound does not travel in a vacuum.

69. A wave generator is used to create a succession of waves. The rate of wave generation is one every 0.33 seconds. The period of these waves is

    A. 2.0 seconds

    B. 1.0 seconds

    C. 0.33 seconds

    D. 3.0 seconds

Answer:

C. 0.33 seconds

The definition of a period is the length of time between wave crests. Therefore, when waves are generated one per 0.33 seconds, that same time (0.33 seconds) is the period. This is consistent only with answer (C). Do not be trapped into calculating the number of waves per second, which might lead you to choose answer (D).

70. In a fission reactor, heavy water

    A. Cools off neutrons to control temperature

    B. Moderates fission reactions

    C. Initiates the reaction chain

    D. Dissolves control rods

Answer:

B. Moderates fission reactions

In a nuclear reactor, heavy water is made up of oxygen atoms with hydrogen atoms called 'deuterium,' which contain two neutrons each. This allows the water to slow down (moderate) the neutrons, without absorbing many of them. This is consistent only with answer (B).

71. Heat transfer by electromagnetic waves is termed

    A. Conduction

    B. Convection

    C. Radiation

    D. Phase Change

Answer:

C. Radiation

To answer this question, recall the different ways that heat is transferred. Conduction is the transfer of heat through direct physical contact and molecules moving and hitting each other. Convection is the transfer of heat via density differences and flow of fluids. Radiation is the transfer of heat via electromagnetic waves (and can occur in a vacuum). Phase Change causes transfer of heat (though not of temperature) in order for the molecules to take their new phase. This is consistent, therefore, only with answer (C).

72. Solids expand when heated because

    A. Molecular motion causes expansion

    B. $PV = nRT$

    C. Magnetic forces stretch the chemical bonds

    D. All material is effectively fluid

Answer:

A. Molecular motion causes expansion

When any material is heated, the heat energy becomes energy of motion for the material's molecules. This increased motion causes the material to expand (or sometimes to change phase). Therefore, the answer is (A). Answer (B) is the ideal gas law, which gives a relationship between temperature, pressure, and volume for gases. Answer (C) is a red herring (misleading answer that is untrue). Answer (D) may or may not be true, but it is not the best answer to this question.

73. Gravitational force at the earth's surface causes

    A. All objects to fall with equal acceleration, ignoring air resistance

    B. Some objects to fall with constant velocity, ignoring air resistance

    C. A kilogram of feathers to float at a given distance above the earth

    D. Aerodynamic objects to accelerate at an increasing rate

Answer:

A. All objects to fall with equal acceleration, ignoring air resistance

Gravity acts to cause equal acceleration on all objects, though our atmosphere causes air resistance that slows some objects more than others. This is consistent only with answer (A). Answer (B) is incorrect, because ignoring air resistance leads to the result of constant acceleration, not zero acceleration. Answer (C) is incorrect because all objects (except tiny ones in which random Brownian motion is more significant than gravity) eventually fall due to gravity. Answer (D) is incorrect because it is not related to the constant acceleration due to gravity.

74. An office building entry ramp uses the principle of which simple machine?

    A. Lever

    B. Pulley

    C. Wedge

    D. Inclined Plane

Answer:

D. Inclined Plane

To answer this question, recall the definitions of the various simple machines. A ramp, which trades a longer traversed distance for a shallower slope, is an example of an Inclined Plane, consistent with answer (D). Levers and Pulleys act to change size and/or direction of an input force, which is not relevant here. Wedges apply the same force over a smaller area, increasing pressure—again, not relevant in this case.

75. The velocity of sound is greatest in

   A. Water

   B. Steel

   C. Alcohol

   D. Air

Answer:

B. Steel

Sound is a longitudinal wave, which means that it shakes its medium in a way that propagates as sound traveling. The speed of sound depends on both elastic modulus and density, but for a comparison of the above choices, the answer is always that sound travels faster through a solid like steel, than through liquids or gases. Thus, the answer is (B).

76. All of the following phenomena are considered "refractive effects" except for

   A. The red shift

   B. Total internal reflection

   C. Lens dependent image formation

   D. Snell's Law

Answer:

A. The red shift

Refractive effects are phenomena that are related to or caused by refraction. The red shift refers to the Doppler Effect as applied to light when galaxies travel away from observers. Total internal reflection is when light is totally reflected in a substance, with no refracted ray into the substance beyond (e.g. in fiber optic cables). It occurs because of the relative indices of refraction in the materials. Lens dependent image formation refers to making images depending on the properties (including index of refraction) of the lens. Snell's Law provides a mathematical relationship for angles of incidence and refraction. Therefore, the only possible answer is (A).

77. Static electricity generation occurs by

    A. Telepathy

    B. Friction

    C. Removal of heat

    D. Evaporation

Answer:

B. Friction

Static electricity occurs because of friction and electric charge build-up. There is no such thing as telepathy, and neither removal of heat nor evaporation are causes of static electricity. Therefore, the only possible answer is (B).

78. The wave phenomenon of polarization applies only to

    A. Longitudinal waves

    B. Transverse waves

    C. Sound

    D. Light

Answer:

B. Transverse waves

To answer this question, recall that polarization is when waves are screened so that they come out aligned in a certain direction. (To illustrate this, take two pairs of polarizing sunglasses, and note the light differences when rotating one lens over another. When the lenses are polarizing perpendicularly, no light gets through.) This applies only to transverse waves, which have wave parts to align. Light can be polarized, but it is not the only wave that can be. Thus, the correct answer is (B).

79. A force is given by the vector 5 N x + 3 N y (where x and y are the unit vectors for the x- and y- axes, respectively). This force is applied to move a 10 kg object 5 m, in the x direction. How much work was done?

    A. 250 J

    B. 400 J

    C. 40 J

    D. 25 J

Answer:

D. 25 J

To find out how much work was done, note that work counts only the force in the direction of motion. Therefore, the only part of the vector that we use is the 5 N in the x-direction. Note, too, that the mass of the object is not relevant in this problem. We use the work equation:
Work = (Force in direction of motion) (Distance moved)
Work = (5 N) (5 m)
Work = 25 J
This is consistent only with answer (D).

80. A satellite is in a circular orbit above the earth. Which statement is false?

    A. An external force causes the satellite to maintain orbit.

    B. The satellite's inertia causes it to maintain orbit.

    C. The satellite is accelerating toward the earth.

    D. The satellite's velocity and acceleration are not in the same direction.

Answer:

B. The satellite's inertia causes it to maintain orbit.
To answer this question, recall that in circular motion, an object's inertia tends to keep it moving straight (tangent to the orbit), so a centripetal force (leading to centripetal acceleration) must be applied. In this case, the centripetal force is gravity due to the earth, which keeps the object in motion. Thus, (A), (C), and (D) are true, and (B) is the only false statement.

www.ingramcontent.com/pod-product-compliance
Lightning Source LLC
Chambersburg PA
CBHW081227090426
42738CB00016B/3216